Climate Politics in Small European States

The characteristics of small states generate multiple and contradictory expectations concerning their climate policies and politics. Do small states perceive themselves as market- and rule-takers, which are largely irrelevant to a global problem, and which must prioritise international competitiveness above climate policy goals? Or do their institutions and their small size foster consensus, coordination, and nimble responses to a changing international scene, allowing them to attain competitive advantages and become climate leaders?

Climate Politics in Small European States examines how the characteristics of small states structure climate politics, both enabling and constraining ambitious climate policies. It contributes to our knowledge of how institutions, actors, and ideas shape climate policy and politics. The volume also contributes to redressing a deficit in the attention given to smaller states in the study of comparative climate politics.

The chapters in this book were originally published as a special issue of *Environmental Politics*.

Neil Carter is Professor of Politics in the Department of Politics at the University of York, UK. He is the author of *The Politics of the Environment* (2018).

Conor Little is Lecturer in the Department of Politics and Public Administration at the University of Limerick, Ireland. His research on climate politics focuses on the policy preferences and policy influence of political parties.

Diarmuid Torney is Associate Professor in the School of Law and Government at Dublin City University, Ireland. He co-founded DCU's MSc in Climate Change: Policy, Media and Society.

Climate Politics in Small European States

Edited by
Neil Carter, Conor Little,
and Diarmuid Torney

LONDON AND NEW YORK

First published 2021
by Routledge
2 Park Square, Milton Park, Abingdon, Oxon OX14 4RN

and by Routledge
52 Vanderbilt Avenue, New York, NY 10017

Routledge is an imprint of the Taylor & Francis Group, an informa business

Introduction, Chapters 1–3 and 5–7 © 2021 Taylor & Francis
Chapter 4 © 2019 Antti Gronow, Tuomas Ylä-Anttila, Marcus Carson and
Christofer Edling. Originally published as Open Access.

British Library Cataloguing in Publication Data
A catalogue record for this book is available from the British Library

ISBN 13: 978-0-367-63996-9

Typeset in MinionPro
by Newgen Publishing UK

Publisher's Note
The publisher accepts responsibility for any inconsistencies that may have arisen
during the conversion of this book from journal articles to book chapters, namely
the inclusion of journal terminology.

Disclaimer
Every effort has been made to contact copyright holders for their permission to
reprint material in this book. The publishers would be grateful to hear from any
copyright holder who is not here acknowledged and will undertake to rectify any
errors or omissions in future editions of this book.

Contents

Citation Information

The chapters in this book were originally published in *Environmental Politics*, volume 28, issue 6 (September 2019). When citing this material, please use the original page numbering for each article, as follows:

Introduction

Climate politics in small European states
Neil Carter, Conor Little and Diarmuid Torney
Environmental Politics, volume 28, issue 6 (September 2019), pp. 981–996

Chapter 1

Does size matter? Comparing the party politics of climate change in Australia and Norway
Fay Madeleine Farstad
Environmental Politics, volume 28, issue 6 (September 2019), pp. 997–1016

Chapter 2

Drivers of political parties' climate policy preferences: lessons from Denmark and Ireland
Robert Ladrech and Conor Little
Environmental Politics, volume 28, issue 6 (September 2019), pp. 1017–1038

Chapter 3

Creative and disruptive elements in Norway's climate policy mix: the small-state perspective
Stefan Ćetković and Jon Birger Skjærseth
Environmental Politics, volume 28, issue 6 (September 2019), pp. 1039–1060

For any permission-related enquiries please visit:
www.tandfonline.com/page/help/permissions

Notes on Contributors

Mikael Skou Andersen, Department of Environmental Science, Aarhus University, Denmark.

Mats Braun, Department of International Relations and European Studies, Metropolitan University Prague, Czech Republic.

Marcus Carson, Stockholm Environment Institute, Sweden.

Neil Carter, Department of Politics, University of York, UK.

Stefan Ćetković, Bavarian School of Public Policy, Technical University of Munich, Germany.

Christofer Edling, Faculty of Social Sciences, Lund University, Sweden.

Fay Madeleine Farstad, CICERO Centre for International Climate Research, Oslo, Norway.

Antti Gronow, Faculty of Social Sciences, University of Helsinki, Finland.

Robert Ladrech, School of Politics, Philosophy, International Relations and Environment, Keele University, UK.

Conor Little, Department of Politics and Public Administration, University of Limerick, Ireland.

Jon Birger Skjærseth, Fridtjof Nansen Institute, Norway.

Diarmuid Torney, School of Law and Government, Dublin City University, Ireland.

Tuomas Ylä-Anttila, Faculty of Social Sciences, University of Helsinki, Finland.

Climate politics in small European states

Neil Carter🆔 , Conor Little🆔 and Diarmuid Torney🆔

The national level has become increasingly important in the study and practice of climate politics. In its move towards a more bottom-up architecture, the Paris Agreement has ushered in a new focus on domestic policy through nationally determined contributions. Scholarly attention has seen a similar shift. Once principally the domain of international relations scholars, the study of climate change politics has paid increasing attention to the determinants of national action (Harrison and Sundstrom 2010, Steinberg and Vandeveer 2012, Cao et al. 2014). Starting from the premise that domestic politics and national governments have a key role to play in climate policymaking, this literature has focused on the politics of national climate legislation (Carter and Jacobs 2014, Lorenzoni and Benson 2014, Fankhauser et al. 2015a, 2015b, Torney 2017, Averchenkova et al. 2018, Wagner and Ylä-Anttila 2018), carbon taxation (Harrison 2010, 2012), and on broader sets of climate policy outputs and outcomes (Christoff and Eckersley 2011, Jensen and Spoon 2011, Compston and Bailey 2012, Bernauer and Böhmelt 2013, Boasson 2013, Lachapelle and Paterson 2013, Tobin 2017).

Although much has been written about the politics of climate change at European and global levels, and with respect to some larger countries such as Germany, the UK, France, the United States, and China, fewer scholarly contributions have focused on small states. Smallness could be defined according to a range of criteria, such as geography or economy, but we define small states in relational terms, that is to say, on the basis of some meaningful criterion in a particular context (Thorhallsson and Wivel 2006). In the context of climate politics and policy, we consider a country's contribution to global greenhouse gas emissions (GHGs) to be a relevant selection criterion. We identify 'small' states in this context as countries accounting for no more than 0.5% of GHGs.[1] Large Annex I countries are the subject of at least five times as many political science, international relations, or public administration publications on climate policy and politics as small Annex I countries. An intra-European comparison of climate policy scholarship on large and small states shows that over twice as many publications feature nine large European states as feature 32 small European states. On average, each of these large European states is the subject of 55 such publications, ranging from 214 (the UK) to 10 (the Ukraine), while the average small state features in eight publications, ranging from 42 (Norway) to 0 in several cases (Web of Science 2018).[2] Thus, relatively little has been written about a large number of states. While this may be understandable, in that individually and collectively they account for a small proportion of the policy problem, it is nonetheless a weakness in the existing literature. These states collectively represent almost one-fifth of European emissions. From the perspective of comparative political science, we are missing opportunities to study and learn lessons from the diverse experiences of climate policy and politics in many small European states. This volume contributes to redressing the deficit in attention given to smaller states in the study of comparative climate politics.

As well as adding to climate politics research by examining understudied cases, we also aim to examine how the characteristics of small states influence climate policy and politics. These characteristics generate multiple, albeit sometimes contradictory, expectations in relation to climate policy and politics. Small states, their governments, and their citizens may see themselves as being irrelevant to GHGs emissions (i.e., a 'drop in the ocean') and thus being without culpability or the means to respond. Being more open to forces of international competition, they may be more susceptible to a 'race to the bottom' on climate policy; their economies may be less diverse than those of larger states and in some cases, this may lead to heavy reliance on GHG-intensive industries. On the other hand, states with small populations may be better equipped to sustain collective action on climate change; under some conditions small European states may be able to influence EU policy initiatives to an extent that is disproportionate to their size; and insofar as they are open economies, they may aim for competitive advantages in new technologies and industries. This volume is motivated by a desire to examine how these contradictory expectations play out in a diverse set of small European states.

Furthermore, our focus on small states is warranted because small state dilemmas are relevant both to small and large countries (Katzenstein 1985). For example, arguments concerning relevance and responsibility for global climate change exist in many larger industrial societies, including larger EU member states and even the United States.

In this introduction, we set out our theoretical expectations regarding climate politics and policy outputs and outcomes in small European states. Drawing on the comparative climate politics literature as well as the literature on the political economy of small states, the next section identifies small state attributes that can both enable and constrain ambitious climate policies. Against this backdrop, we go on to highlight the empirical findings of the various contributions to the volume and explore the proposition that small states share certain distinctive characteristics that are relevant to climate politics and policy.

Constraints and opportunities for small states

The literature on small states that has developed around the work of Peter Katzenstein (1985) has implications for both small state policy in the face of international challenges and the domestic politics of small states, not least in relation to internationalised issues that often present tensions with market forces. These implications include the degree of consensus in their climate politics and conflicting expectations concerning small state leadership and laggardship in climate policy outputs.

Consensus-oriented politics and convergent preferences in the face of external pressures are strong themes in research on small states. Katzenstein argued that small states are vulnerable due to geopolitical factors and their dependence on trade. This vulnerability provides the motivation, and their small size and corporatist institutions provide the means, for coordination among policymakers and other important actors, making for consensus-oriented, 'low voltage' or even 'unitary' politics, which 'makes political conflicts over basic political choices illegitimate' (Katzenstein 1985, pp. 32, 208–209). Consensus-oriented politics has in part been attributed to corporatist institutions but has also been linked to smallness *per se*. Keating (2015, pp. 16–17) emphasises the short lines of communication, a shared sense of purpose, and trust among governing elites as conditions that facilitate consensus in small states. Consensus, in turn, may facilitate policy consistency over time (e.g., Ingebritsen 2002), which can be conducive to stronger policy action on climate change due to the time horizons over which costs and benefits of such action play out. The apparent advantages for climate leadership of political consensus are reflected in both empirical research and policy recommendations (e.g., Brennan and Curtin 2008, p. 56, Flagg 2015). On the other hand, polarisation may lead to greater political constraints on decision-makers and greater inconsistency and policy uncertainty over time. This line of argument intersects with findings on climate policy that highlights the role of coordination in producing effective policy (Christoff and Eckersley 2011, Lachapelle and Paterson 2013, Ćetković and Buzogány 2016; see also Ingebritsen 2010).

The implications for *climate policy leadership* that can be drawn from the literature on small states are somewhat ambiguous. On the one hand, small state attributes may deter governments from climate policy leadership. Fundamentally, citizens and elites in small states may see their governments' domestic policy decisions as being practically irrelevant to global

GHG emissions (Thorhallsson and Wivel 2006, p. 654), thus undermining the case for leadership. In addition, small states rely heavily on external markets; they are often policy- and market-takers; and they have a lower capacity to resist external pressure (Katzenstein 1985, Keating 2015). This too may lead them to eschew leadership positions on climate policy. Further, ambitious climate policies may be dampened by the domestic politics of small states: they may be prone to risk aversion or group-think (Keating and Harvey 2014: 62, Carolan 2017). Small state economies may be particularly likely to be heavily reliant on individual sectors, and their policy processes may be prone to 'sectoral corporatism' (Lehmbruch 1984), constraining decision-makers and insulating policymaking from wider societal debates.

On the other hand, the distinctive characteristics of small states, particularly their responsiveness to market and regulatory developments, may lead them to act as innovators or test beds for new policy responses to climate change. The increasingly polycentric nature of climate governance, characterised by multiple more or less independent locations of decision-making, places a premium on innovations in climate governance (Jordan and Huitema 2014, Jordan *et al.* 2018). Small states may be particularly well equipped to act as climate policy innovators, with successful innovations spreading out from their initial source through learning and other transfer mechanisms. Indeed, this focus on experimentation has been a hallmark of the literature on polycentric climate governance (Ostrom 2010, 2012, Dorsch and Flachsland 2017), although Jordan *et al.* (2015) caution that this optimistic take on polycentric governance rests on untested assumptions about the diffusion and performance of governance innovations. Small states may also be likely destinations to which innovations diffuse: small states are more open to international forces of coercion and competition, both from competitiveness pressures and from the impetus to develop comparative advantage, and may also be more likely to engage in emulation and learning (Katzenstein 2003, p.18, Simmons *et al.* 2006). Small state vulnerability may contribute to 'the greening of capitalism' (Ingebritsen 2010, p. 362).

There is some empirical evidence for small state leadership in climate policy. For example, small states were early adopters of carbon taxes: in the 1990s and 2000s, nearly all of the European countries that implemented carbon taxes at the national level were small states. Small European states have also tended to set their carbon prices at higher levels (World Bank 2018; see also Andersen 2019). Taking a broader view of climate policy outputs and outcomes, small European states have tended to achieve somewhat higher values in the Climate Change Performance Index (CCPI) than larger states in recent years (Burck *et al.* 2017; Figure 1). Moreover, public opinion in small European states has tended to indicate more concern about climate change and related environmental and energy issues than

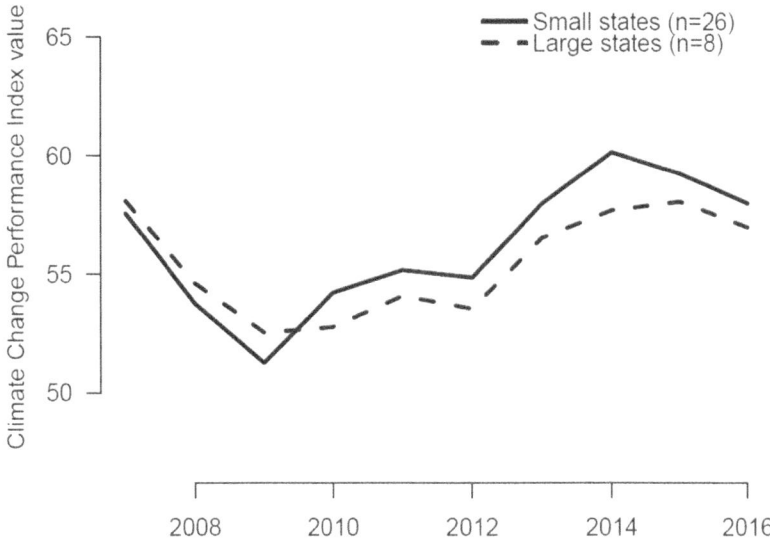

Data: Burck et al. (2017) and previous years' Climate Change Performance Indexes.

Figure 1. Mean values of small and large European states in the Climate Change Performance Index.

public opinion in larger states (European Commission 2018). Figure 2 illustrates that difference; a similar mean difference between small and large European states of approximately two percentage points is evident in responses to questions about the most important problem facing the EU and facing the respondent personally.[3]

Each of the tendencies set out thus far has identified potential commonalities in the climate politics in small states. However, *diversity* among small states is widely acknowledged (e.g., Katzenstein 1985, 2003, p. 11, Thorhallsson and Wivel 2006, p. 655, Antunes and Loughlin 2018, p. 5). Small European states include (neo-)corporatist states and others (Thorhallsson and Kattel 2013), social investment and market competition states (Keating and Harvey 2014), and Coordinated and Liberal Market Economies (Hall and Soskice 2001). While they are all parliamentary systems, their institutions for representative politics also differ in important ways. These differences can be expected to influence how small states respond to global policy problems and they have been associated – with varying degrees of confidence and empirical evidence – with different approaches to climate policy (Christoff and Eckersley 2011, Lachapelle and Paterson 2013, Ćetković and Buzogány 2016). It may be that several of the features of climate politics discussed above are contingent on these institutional features, and not on state smallness alone.

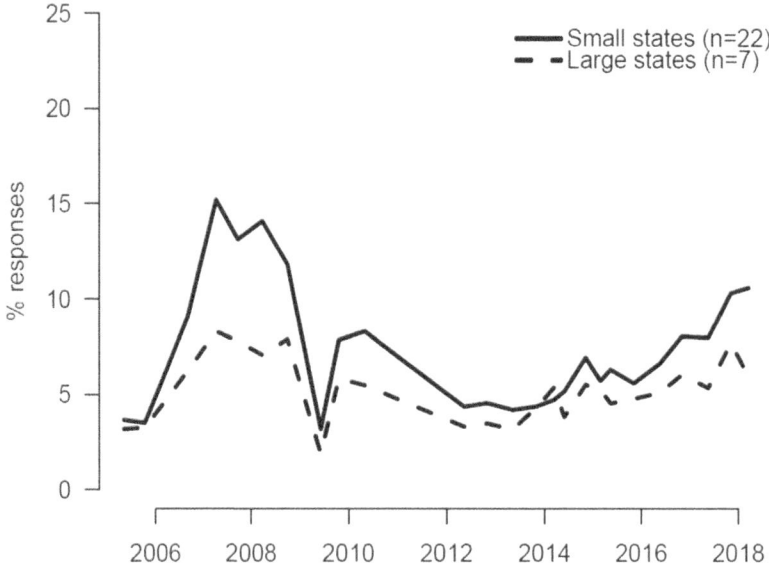

Figure 2. Public concern about the environment, climate change, and energy in small and large European states.

The contributions to this volume

The contributions to this volume explore how the themes developed above, including consensus (and polarisation) in domestic climate politics and leadership and laggardship, as well as the role of corporatist institutions, play out in different small state contexts. The first two contributions examine *consensus and polarisation* on climate policy among political parties. Farstad (2019) compares the party politics of climate change in a large state (Australia) with a small European state (Norway), examining the possible effects of state size, while Ladrech and Little (2019) examine differences between two small European states (Denmark and Ireland) and variations within those states, emphasising diversity in the climate politics of small European states.

Farstad has a positive story to tell about small states and party consensus on climate policy. She asks whether and how state size has an effect on party consensus on climate policy. She compares a large state (Australia) and a small state (Norway) that are similar on several dimensions: both are major fossil fuel exporters, the cost of abatement in both countries is high, and both of these wealthy countries were relatively unscathed by the global economic crisis. Yet, the party politics of climate change in Australia has become polarised, while there has been considerable consensus in Norway. On the basis of these case studies drawing on 44 interviews with policy practitioners, she concludes that geographic size conditions party behaviour on climate change, contributing – to a limited extent – to explanations of consensus (polarisation). Geographic size operates by underpinning inter-regional variation in economic interests, which further interacts with party ideology to accentuate polarisation. She also finds some evidence for the proposition that smallness facilitates shorter lines of communication and greater unity among elites. However, institutions matter more than size in both countries: while acknowledging a range of institutions that may con-tribute to consensus (polarisation) on climate change between and within parties, Farstad identifies corporatism as a key factor that conditions party preferences and behaviour on climate change by engendering coordination and consensus among interest groups and ENGOs.

Ladrech and Little begin from the observation that not only do parties' preferences on climate policy differ but they also appear to depend on parties' positions on the left-right spectrum. They examine the roles of three factors in driving parties' climate policy preferences and left-right differences in those preferences: public opinion, party competition, and pre-existing policy preferences on traditional policy issues. Denmark and Ireland appear typical of the observed relationship between left-right pol-itics and climate politics: Denmark is somewhat polarised in both its left-right politics and its climate politics (albeit with considerable variation over time), while in Ireland party preferences are more convergent. They find that public opinion and party competition had strong effects in shaping party preferences, but that these effects were uneven across the left-right spectrum in each country, with typically greater incentives to develop strong climate policy preferences on the left. The role of public opinion is exemplified both by the increased vote-seeking incentives to address the issue in both countries in the late 2000s and by how low public concern in Ireland has allowed parties to maintain relatively weak climate policy preferences. Party competition has driven accommodative behaviour towards successful issue-owners. They add to existing research that high-lights the competitive dynamic that can occur between larger parties on climate policy (Carter and Jacobs 2014) by showing the – perhaps more typical – constraining effects of mainstream competition on climate policy.

They also show that parties' existing policy preferences on traditional issues, especially on economic issues, have a particularly important role to play in shaping parties' climate policy preferences, which helps to explain observed differences in climate policy preferences between parties of the left and right. The drivers of parties' preferences on climate policy have much in common with their incentives and constraints in other policy domains, and there is little to suggest a generic imprint of 'small stateness' on the party politics of climate change. While the study detects some small-state-specific drivers – such as perceptions of national irrelevance to global greenhouse gas emissions among some politicians in Ireland – the case of Denmark shows that these are not universal or inevitable.

The next two contributions focus on *corporatism* in Nordic states. Ćetković and Skjærseth (2019) analyse the creative and disruptive elements of Norway's climate policy mix, distinguishing three main phases in its development. The first phase (1989–1995), termed 'symbolic destruction', emphasised cost-efficient stabilisation of domestic emissions. In the second phase (the mid-1990s–2008), attention shifted towards incremental improvements in environmental efficiency. The third phase (since 2008) has been characterised by increasing efforts towards emissions reductions at home combined with intensive global climate diplomacy. However, they argue, the actual impact of these cumulative changes on climate policy has been limited. The authors demonstrate how the mixed efforts for mitigating climate change domestically and high activity in the global climate governance regime have been tied to Norway's political economy and foreign policy strategy as a small, open, social-investment economy. Their findings support the theoretical proposition that small corporatist economies are successful in incremental long-term adaptation but prone to ignore larger structural problems. Whereas Norway has achieved a remarkably stable climate policy consensus and has continuously encouraged the environmental improvements in the oil and gas sector, it has failed to formulate a plan for phasing out the oil and gas extraction and reduce the country's economic dependence on oil and gas exports.

Although Norway's climate policy can partly be described as symbolic, for instance, the modest CO_2 tax, pressure to live up to its 'green' reputation has motivated Norway to undertake some more structural reforms such as the extensive promotion of electric vehicles. These progressive measures indicate that small social-investment economies like Norway may be more inclined to increase their climate policy efforts in the face of external pressure as they are more dependent on stable international agreements and a progressive self-image than larger and more liberal-market-oriented economies such as Australia and Canada (see Cass 2008).

Gronow *et al.* (2019) compare two small Nordic states – Sweden and Finland – that are similar in many respects but differ with regard to both climate change policy outputs and outcomes. They seek to resolve the debate between those who argue that corporatism is associated with more ambitious environmental policies (for example, Christoff and Eckersley 2011) and those who maintain that corporatism can hinder rather than promote ambitious environmental policy (Dryzek *et al.* 2003). They suggest that, rather than treating corporatism either as a dichotomous macro-structural variable or a continuum on which all countries can be placed, the institution of corporatism should be divided into three components: inclusiveness, consensualism and strength of tripartite organisations. Gronow *et al.* maintain that although inclusiveness and consensualism are indeed conducive to ambitious environmental policy, tripartite strength may have the opposite effect. Their empirical findings suggest that Sweden and Finland are both characterised by relatively inclusive and consensual policymaking systems, but that in Finland, NGOs are less influential than tripartite organisations, whereas in Sweden the reverse is true. Finnish tripartite organisations are influential and occupy important positions in a resourceful coalition well linked to the government. This contrasts with Sweden and appears to be the most important difference in the climate policy networks between the two countries.

The remaining three contributions address various aspects of *leadership and laggardship*, without dispensing with the themes of political consensus and corporatism. Andersen (2019) builds on the observation that small states have been front-runners in the adoption of carbon taxes to examine seven cases of the adoption of that policy instrument from a range of different contexts within the EU: the Nordic states (Sweden, Finland, and Denmark), central and eastern Europe (Slovenia and Estonia), and two 'cohesion' countries (Ireland and Portugal). Drawing on the concept of 'policy styles', he argues that, despite their differences, these cases are characterised by similarities that helped to make their contexts conducive to the adoption of carbon taxes. Two elements are central to his argument. First, five of the seven states were characterised by a deep-rooted national policy style involving routines of interest coordination applied to fiscal policy that allowed a proactive approach to taxation policy. In contrast to the contributions that precede it in this volume, neo-corporatist structures play a less prominent role than expected; rather, cultural norms concerning consultation and coordination are the common denominator. Andersen observes that these countries frequently introduced carbon taxes as part of broader fiscal reforms, allowing those who were disadvantaged by the introduction of the tax to be compensated elsewhere. Second, their consensus-orientation was reinforced by another characteristic typical of small states: their proportional electoral systems, which broadened parliamentary representation, and which in some cases allowed environmentally oriented parties to influence agenda-setting or the adoption of a carbon tax (e.g., in Denmark, Finland, and Ireland).

Braun (2019) finds evidence of climate leadership in his study of the role of the Czech Republic in EU climate change negotiations, including its coordination with the other members of the so-called Visegrad Group (V4, also including Poland, Hungary and Slovakia). The role of the central and eastern European member states in EU climate and energy policy has received some coverage in the scholarly literature, but less attention has been paid to the role the smaller Visegrad countries play in the development of EU climate policy. Braun argues that institutionalised cooperation within the V4 was a crucial component of their approach to the negotiations of the EU 2030 framework. The V4 countries managed to coordinate their positions during the negotiations on the 2030 framework, and several of the group's initial demands were reflected in the final package agreed upon by the European Council in October 2014. He utilises the concepts of rhetorical action and socialisation, arguing that the V4 group's appeal to economic fairness corresponds to the idea of rhetorical action since the demands for a burden-sharing agreement reflect a shared EU norm of economic solidarity. At the same time, the group's approach has been developed as a response to and in conflict with ideas of ecological modernisation that legitimise EU climate policy. The cooperation and frequent meetings between representatives of the V4 provide a potential platform for a process of socialisation that in this case reinforces norms that contradict the Europeanisation process. He concludes that the success of the Visegrad cooperation in EU climate and energy negotiations suggests that researchers should take this form of international cooperation seriously despite the limited levels of formal institution building within the V4.

Torney (2019), however, finds that domestic factors can constrain the positive impact of international pressure. He investigates the adoption of framework climate laws in two small European states, Ireland and Finland, both of which introduced national climate laws in 2015. He tests two propositions: first, that international pressure to act on climate change pushes policymakers to develop climate policies that draw on experiences from other jurisdictions; and second, that interest group mobilisation and political contestation over climate policy will cause states to adopt weaker, more symbolic versions of pioneering policies in other jurisdictions.

In respect of the first proposition, the UK's 2008 Climate Change Act was a source of inspiration in the early stages of the legislative process, driven by international pressure to be seen to be acting on climate change. The lead-up, first to COP15 and later to COP21, created conditions in which policymakers wanted to be seen to be doing *something*. In respect of the second proposition, however, strong societal conflict led both Ireland and Finland to adopt weaker climate laws. Domestic interest groups mobilised to remove the most pioneering and ambitious parts of the UK model from legislative proposals, with the preferences of more powerful actors trumping NGO voices pushing for stronger legislation. In this respect, two key elements that facilitated passage of the UK Climate Change Act (Carter and Childs 2018) were absent in Ireland and Finland: support from large elements of business, and cross-party consensus on the need for strong climate legislation. Thus, expectations regarding smallness are only partially borne out. Competitiveness concerns underpinned by both countries being small, open economies were a strong constraint on the adoption of more ambitious legislation. On the other hand, consensus-based politics (in the case of Finland) did not facilitate the adoption of comparatively ambitious legislation (see also Gronow *et al.* 2019 – this volume).

Overall, the cases and perspectives explored in this volume illustrate the diversity that exists with respect to the expectations set out above. While consensus on climate policy is a prominent feature in some of the countries examined, such as Norway and Sweden, several other cases are characterised by significant political partisanship or deep-rooted societal contestation, including in Denmark, Ireland, Finland and the Czech Republic, among others. Corporatism – or its absence – is seen to play a role in shaping climate policy outcomes, but a nuanced picture emerges across our cases. Corporatism shapes outcomes most prominently in the Nordic countries, but even here there is significant variation, with stronger climate policy outcomes in Norway and Sweden than in Finland. Political institutions matter in other ways as well, including electoral institutions and informal institutional arrangements at the supranational level such as the Visegrad 4 group in which the Czech Republic is embedded. Depending on the constellation of actors and interests, small states can play varying roles in regional and global climate politics, including as leaders (e.g., Norway and Sweden), laggards (e.g., Ireland and the Czech Republic) or somewhere in between (e.g., Finland). They can be innovators and sources of policy diffusion, for example, on carbon taxes, but they can also be at the receiving end of diffusion processes and may, along the way, modify or weaken policy innovations, as has been the case with climate legislation. The overall picture that emerges is one of diversity in climate politics and policy across small European states driven by a range of factors in which size has an occasional role.

This volume contributes to redressing the balance of scholarly attention in favour of small states. It touches on several least-studied states, not least in its inclusion of the case of carbon taxes in Portugal, Estonia, and Slovenia (Andersen), and climate policy development in the Czech Republic in the context of the broader V4 group (Braun). However, it also reinforces some existing trends that favour some small states over others. It covers small states that have received more attention, such as the Nordic states. Nonetheless, our observations in this Introduction concerning the uneven distribution of scholarly attention to climate policy in small states and the volume's findings, which emphasise diversity among these states, suggest that even within Europe there is considerable further work to be done to expand the empirical scope of the comparative study of climate policy and politics. In particular, future research could focus more attention on eastern and southern European small states. The net could also be cast wider to consider small states beyond Europe. Doing so would enable exploration of the mediating role played by EU membership in the climate politics of small European states. Furthermore, and building on the observations of Farstad (2019), more research is needed to unpack the relationship between state size defined in terms of emissions, geography, and political institutions. Finally, with one exception (Farstad 2019), this volume has focussed on comparisons among small states. Future work on the role of state size in climate politics may focus on large-small comparisons. The diversity among small states illustrated in this volume, however, suggests that any effects of size sought by such studies are likely to be conditional and complex.

Notes

1. Annex I countries that contributed less than 0.5% to global greenhouse gas emissions in 2014 (taking account of land use change and forestry) were: Austria, Belarus, Belgium, Bulgaria, Croatia, Cyprus, Czech Republic, Denmark, Estonia, Finland, Greece, Hungary, Iceland, Ireland, Latvia, Liechtenstein, Lithuania, Luxembourg, Malta, Monaco, Netherlands, New Zealand, Norway, Portugal, Romania, Slovakia, Slovenia, Sweden, and Switzerland. Annex I countries above this threshold (excluding the European Union) were Australia, Canada, France, Germany, Italy, Japan, Poland, Russia, Spain, Turkey, Ukraine, the United Kingdom, and the United States (CAIT-WRI 2017).
2. Searches for 'climate change' and cognate terms ('global warming', 'climate policy', 'climate politics', and 'greenhouse gas*') were carried out in July and August 2018 in the titles, keywords, and abstracts of publications classified as political science, international relations, or public administration in the *Web of Science*. Variants of country names (e.g., UK, United Kingdom, Britain) were used where necessary.
3. Country selection is based on data availability in the CCPI and Eurobarometer. The countries covered by the Eurobarometer data in Figure 2 are all those for which data is available from 2005 to 2018; these are the EU-28 in 2018 and Turkey. Figure 1 covers the same countries, as well as Belarus, Norway, Switzerland, and the Ukraine. Russia is excluded, as it is both conceptually marginal to 'Europe' and empirically outlying as a climate laggard. One-tailed t-tests on the pooled CCPI observations indicate that the difference between small and large European states illustrated in Figure 1 is not statistically significant ($n=$ 373, $p=$ 0.19); however, they indicate that differences in public opinion between large and small state observations are statistically significant at <0.01 (n ranges from 450 to 724).

Acknowledgments

Early drafts of the contributions to this special issue were presented at a workshop in June 2016 hosted by Dublin City University and supported by the Irish Environmental Protection Agency, the University Association for Contemporary European Studies, the Irish Association for Contemporary European Studies, and the DCU School of Law and Government. Diarmuid Torney thanks Dublin City University for support from the Faculty of Humanities and Social Sciences Journal Publication Scheme. We are very grateful to Louise Fitzgerald for extensive editorial assistance and to the referees as well as to the journal editor-in-chief, Chris Rootes, for their very helpful comments. The usual disclaimer applies.

Disclosure statement

No potential conflict of interest was reported by the authors.

Funding

This work was supported by the Irish Environmental Protection Agency, the University Association for Contemporary European Studies, the Irish Association for Contemporary European Studies, and the DCU School of Law and Government.

ORCID

Neil Carter (iD) http://orcid.org/0000-0003-3378-8773
Conor Little (iD) http://orcid.org/0000-0001-5510-3195
Diarmuid Torney (iD) http://orcid.org/0000-0003-4156-9044

References

Andersen, M.S., 2019. The politics of carbon taxation: how varieties of policy style matter. *Environmental Politics*, 28, 6.

Antunes, S. and Loughlin, J., 2018. The European Union, subnational mobilization and state rescaling in small unitary states: a comparative analysis. *Regional & Federal Studies*. 1–16. online before print. doi:10.1080/13597566.2018.1512974

Averchenkova, A., Fankhauser, S., and Nachmany, M., eds, 2018. *Trends in climate change legislation*. Cheltenham: Edward Elgar.

Bernauer, T. and Böhmelt, T., 2013. National climate policies in international comparison: the climate change cooperation index. *Environmental Science & Policy*, 25, 196–206. doi:10.1016/j.envsci.2012.09.007

Boasson, E. 2013. *National climate policy ambitiousness: a comparative study of Denmark, France, Germany, Norway, Sweden and the UK*. Oslo: Cicero Available from: http://www.cicero.uio.no/media/10128.pdf. [accessed 16 December 2013].

Braun, M., 2019. The Czech Republic's approach to the EU 2030 climate and energy framework. *Environmental Politics*, 28 (6).

Brennan, P. and Curtin, J., 10 January 2008. *Climate change challenge- strategic issues, options and implications for Ireland*. Dublin: Institute for International and European Affairs.

Burck, J., *et al.*, 2017. *The climate change performance index: results 2018*. Bonn/Berlin/Brussels: Germanwatch/CAN Europe.

CAIT-WRI, 2017. CAIT climate data explorer. World Resources Institute. Available from: http://cait.wri.org [Accessed5 June 2019].

Cao, X., *et al.*, 2014. Research frontiers in comparative and international environmental politics an introduction. *Comparative Political Studies*, 47 (3), 291–308. doi:10.1177/0010414013509567

Carolan, E., 2017. Are small states susceptible to groupthink? Lessons for institutional design. *European Political Science*, 16 (3), 383–399. doi:10.1057/s41304-016-0003-9

Carter, N. and And Childs, M., 2018. Friends of the Earth as a policy entrepreneur: 'The Big Ask' campaign for a UK climate change act. *Environmental Politics*, 27 (6), 994–1018. doi:10.1080/09644016.2017.1368151

Carter, N. and Jacobs, M., 2014. Explaining radical policy change: the case of climate change and energy policy under the British Labour government 2006–10. *Public Administration*, 92 (1), 125–141. doi:10.1111/padm.2014.92.issue-1

Cass, L.R., 2008. A climate of obstinacy: symbolic politics in Australian and Canadian policy. *Cambridge Review of International Affairs*, 21 (4), 465–482. doi:10.1080/09557570802452763

Ćetković, S. and Buzogány, A., 2016. Varieties of capitalism and clean energy transitions in the European Union: when renewable energy hits different economic logics. *Climate Policy*, 16 (5), 642–657. doi:10.1080/14693062.2015.1135778

Ćetković, S. and Skjærseth, J.B., 2019. Creative and disruptive elements in the Norway's climate policy mix: the small-state perspective. *Environmental Politics*, 28 (6).

Christoff, P. and Eckersley, R., 2011. Comparing State Responses. *In*: J.S. Dryzek, R. B. Norgaard, and D. Schlosberg, eds. *Oxford handbook of climate change and society*. Oxford: Oxford University Press, 431–448.

Compston, H. and Bailey, I. 2012. *Climate clever: how governments can tackle climate change (and still win elections)*. Abingdon: Routledge Available from: http://orca.cf.ac.uk/23116/. [Accessed 18 September 2013].

Dorsch, M.J. and Flachsland, C., 2017. A polycentric approach to global climate governance. *Global Environmental Politics*, 17 (2), 45–64.

Dryzek, J., *et al.*, 2003. *Green states and social movements*. Oxford: Oxford University Press.

European Commission, 2018. *Regular Eurobarometer surveys*. Available from: http://ec.europa.eu/commmfrontoffice/publicopinion/index.cfm [Accessed 10 July 2018].

Fankhauser, S., Gennaioli, C., and Collins, M., 2015a. Do international factors influence the passage of climate change legislation? *Climate Policy*, 16 (3), 318–331. doi:10.1080/14693062.2014.1000814

Fankhauser, S., Gennaioli, C., and Collins, M., 2015b. The political economy of passing climate change legislation: evidence from a survey. *Global Environmental Change*, 35, 52–61. doi:10.1016/j.gloenvcha.2015.08.008

Farstad, F., 2019. Does size matter? Comparing the party politics of climate change in Australia and Norway. *Environmental Politics*, 28 (6).

Flagg, J.A., 2015. Aiming for zero: what makes nations adopt carbon neutral pledges? *Environmental Sociology*, 1 (3), 202–212. doi:10.1080/23251042.2015.1041213

Gronow, A., *et al.*, 2019. Divergent neighbors: corporatism and climate policy networks in Finland and Sweden. *Environmental Politics*, 28 (6).

Hall, P. and Soskice, D., eds., 2001. *Varieties of capitalism: the institutional foundations of comparative advantage*. Oxford: Oxford University Press.

Harrison, K., 2010. The comparative politics of carbon taxation. *Annual Review of Law and Social Science*, 6 (1), 507–529. doi:10.1146/annurev. lawsocsci.093008.131545

Harrison, K., 2012. A tale of two taxes: the fate of environmental tax reform in Canada. *Review of Policy Research*, 29 (3), 383–407. doi:10.1111/j.1541-1338.2012.00565.x

Harrison, K. and Sundstrom, L., 2010. *Global commons, domestic decisions: the comparative politics of climate change*. Cambridge, MA: MIT Press.

Ingebritsen, C., 2002. Norm entrepreneurs. Scandinavia's role in world politics. *Cooperation and Conflict*, 37 (1), 11–23. doi:10.1177/0010836702037001689

Ingebritsen, C., 2010. Katzenstein's legacy 25 years after: small states in world markets. *European Political Science*, 9 (3), 359–364. doi:10.1057/eps.2010.23

Jensen, C. and Spoon, -J.-J., 2011. Testing the 'Party Matters' thesis: explaining progress towards Kyoto protocol targets. *Political Studies*, 59 (1), 99–115. doi:10.1111/j.1467-9248.2010.00852.x

Jordan, A., *et al.*, 2015. Emergence of polycentric climate governance and its future prospects. *Nature Climate Change*, 5, 977–982. doi:10.1038/nclimate2725

Jordan, A., *et al.*, eds, 2018. *Governing climate change: polycentricity in action?* Cambridge: Cambridge University Press.

Jordan, A. and Huitema, D., 2014. Innovations in climate policy: the politics of invention, diffusion, and evaluation. *Environmental Politics*, 23 (5), 715–734. doi:10.1080/09644016.2014.923614

Katzenstein, P., 1985. *Small states in world markets. industrial policy in Europe*. Cornell studies in political economy. New York: Cornell University Press.

Katzenstein, P., 2003. Small states and small states revisited. *New Political Economy*, 8 (1), 9–30. doi:10.1080/1356346032000078705

Keating, M., 2015. The political economy of small states in Europe. *In*: H. Baldersheim and M. Keating, eds. *Small states in the modern world : vulnerabilities and opportunities*. Cheltenham: Edward Elgar, 1–19.

Keating, M. and Harvey, M., 2014. The political economy of small European states: and lessons for Scotland. *National Institute Economic Review*, 227 (1), R54–R66. doi:10.1177/002795011422700107

Lachapelle, E. and Paterson, M., 2013. Drivers of national climate policy. *Climate Policy*, 13 (5), 547–571. doi:10.1080/14693062.2013.811333

Ladrech, R. and Little, C., 2019. Drivers of political parties' climate policy preferences: lessons from Denmark and Ireland. *Environmental Politics*, 28, 6.

Lehmbruch, G., 1984. Concertation and the structure of corporatist networks. *In*: J. Goldthorpe, ed. *Order and conflict in contemporary capitalism*. Oxford: Clarendon Press, 60–80.

Lorenzoni, I. and Benson, D., 2014. Radical institutional change in environmental governance: explaining the origins of the UK Climate Change Act 2008 through discursive and streams perspectives. *Global Environmental Change*, 29, 10–21. doi:10.1016/j.gloenvcha.2014.07.011

Ostrom, E., 2010. Polycentric systems for coping with collective action and global environmental change. *Global Environmental Change*, 20 (4), 550–557.

Ostrom, E., 2012. Nested externalities and polycentric institutions: must we wait for global solutions to climate change before taking actions at other scales? *Economic Theory*, 49 (2), 353–369.

Simmons, B., Dobbin, F., and Garrett, G., 2006. Introduction: the international diffusion of liberalism. *International Organization*, 60 (04), 781–810. doi:10.1017/S0020818306060267

Steinberg, P. and Vandeveer, S., 2012. *Comparative environmental politics: theory, practice, and prospects*. Cambridge, MA: MIT Press.

Thorhallsson, B. and Kattel, R., 2013. Neo-liberal small states and economic crisis: lessons for democratic corporatism. *Journal of Baltic Studies*, 44 (1), 83–103. doi:10.1080/01629778.2012.719306

Thorhallsson, B. and Wivel, A., 2006. Small states in the European Union: what do we know and what would we like to know? *Cambridge Review of International Affairs*, 19 (4), 651–668. doi:10.1080/09557570601003502

Tobin, P., 2017. Leaders and laggards: climate policy ambition in developed states. *Global Environmental Politics*, 17 (4), 28–47. doi:10.1162/GLEP_a_00433

Torney, D., 2017. If at first you don't succeed: the development of climate change legislation in Ireland. *Irish Political Studies*, 32 (2), 247–267. doi:10.1080/07907184.2017.1299134

Torney, D., 2019. Climate laws in small European states: symbolic legislation and limits of diffusion in Ireland and Finland. *Environmental Politics*, 28, 6.

Wagner, P. and Ylä-Anttila, T., 2018. Who got their way? Advocacy coalitions and the Irish climate change law. *Environmental Politics*, 27 (5), 872–891. doi:10.1080/09644016.2018.1458406

Web of Science, 2018. *Web of Science*. Available from: https://webofknowledge.com/ . [Accessed 13 July and 2 August 2018].

World Bank, 2018. *State and trends of carbon pricing 2018*. May. Washington DC: World Bank.

Does size matter? Comparing the party politics of climate change in Australia and Norway

Fay Madeleine Farstad

ABSTRACT

The implications of state size for the party politics of climate change are examined, and in particular its effect in facilitating or impeding cross-party consensus on the issue. This issue is explored through an in-depth, qualitative comparison of Australia and Norway, which are shown to be comparable in important respects yet differ in terms of their size and climate politics. Original primary data is presented from 44 interviews with policymakers and policy-shapers in both countries, which shows that, to the limited extent that state size moderates the parties' behaviour, it is the countries' geographical – not economic – size that matters. Institutional factors are found to play a more significant role, however, and the corporatist features of state-business coop-eration, strong ENGOs and compensatory welfare arrangements are high-lighted as particularly important.

Introduction

Although the literature seeking to explain variation in countries' climate change policies and ambition has been developing for some time (e.g. Lachapelle and Paterson 2013, Fankhauser *et al.* 2015), there are important omissions from this body of scholarship. First, the role of political parties has been largely neglected (for exceptions see Båtstrand 2014, 2015, Carter *et al.* 2017, Farstad 2017). Second, although various state features have been examined for their relevance in explaining climate policy variation, 'size' has rarely been one of them. Here, I help to fill both these gaps in the literature, and also provide a significant contribution by examining how these two features interlink, i.e. analysing the implications of size for the party politics of climate change. These are important gaps to fill. Political parties lie at the heart of climate change politics, as party competition significantly shapes government policy and national governments are cen-tral to policymaking on climate change. The critical role of political parties for reaching national climate change targets was noted by Jensen and Spoon

(2011). Furthermore, political parties are important in ensuring the stability and ultimate success of climate policies. Given the long-term character of climate change, commensurately long-term investments are needed. Such investments require continued, and thus bi-partisan, support for their survival and success. Countries with cross-party consensus on climate change often experience more stable and ambitious climate policies (such as the UK and Norway). In contrast, in countries such as Australia, Canada and the US, parties have polarised over the issue, with detrimental effects on climate policy and investment security (Tranter 2013, Dunlap *et al.* 2016). As the countries experiencing party polarisation on climate change tend to be large, and the countries experiencing cross-party consensus are generally small, size seems intuitively to play a critical role in shaping cross-party agreement on climate change, and consequently stable and successful climate policy. It is thus timely that the effect of state size is examined. If state size is revealed to significantly shape the party politics of climate change, this knowledge could be used when designing or re-designing political institutions to improve climate change outcomes.

To test the assumption that size matters for the party politics of climate change this article compares two countries with significant similarities yet differing sizes and climate politics. Australia and Norway are two developed and major fossil fuel exporters, but with smaller Norway experiencing cross-party consensus on its ambitious climate policies whereas large Australia has seen the major parties polarise over climate change, significantly weakening its national policy ambition. The countries are compared through analysing secondary literature and primary data from 44 interviews with policymakers and policy-shapers in both countries.

The first section outlines the theoretical underpinnings of the investigation, namely the comparative climate policy, party politics and state size literature, before the second section outlines and justifies the cases for comparison. The third section reviews the interview findings, before the implications of these for the state size hypothesis and future research are discussed. I argue that the 'size matters' hypothesis finds somewhat limited support in the two case studies, with institutional features playing a more significant role in shaping the party politics of climate change. In particular, I highlight the benefits of corporatist features in facilitating cross-party agreement on climate change. To the limited extent that state size moderates the parties' behaviour on the issue of climate change, it is the countries' geographical – not economic – size that matters.

Variation in climate policy: state size and party politics

In the comparative climate policy literature various factors have been identified as relevant to explaining the variation in states' climate policies. Countries with higher levels of economic prosperity (Neumayer 2002a, Liefferink *et al.* 2009) and democratic quality (Neumayer 2002b, Lachapelle and Paterson 2013) tend to have more ambitious climate change policies, as being wealthy frees up capacity to care about climate change and democracy enables citizens to put pressure on governments to act. Similarly, countries with proportional electoral systems usually have more ambitious climate policies (Scruggs 1999, Harrison and Sundstrom 2010, Spoon *et al.* 2014), as such systems make it easier for small (e.g. green) parties to enter parliament, amplifying the voices of the minority of voters for whom climate change is a priority and increasing party competition on the issue. Coordinated or corporatist economies also tend to have stronger climate policies (Scruggs 1999, Lachapelle and Paterson 2013, Ćetković and Buzogány 2016) as the cooperative relations between state and businesses enable the state to transform the interests and practices of businesses, and the increasing involvement of green groups in such relations counterbalances the potential negative influence of businesses. In contrast, liberal market or pluralist economies usually experience worse environmental outcomes as the competition between interest groups for government access tends to favour wealthier and better-organised business lobbies (Griffiths *et al.* 2007, Bernhagen 2008). Further, countries with high levels of fossil fuel dependency generally have weaker climate policies (Lachapelle and Paterson 2013, Fankhauser *et al.* 2015), reflecting the entrenched power of fossil fuel interests in such countries and problems of 'carbon lock-in' (Unruh 2000). Countries with numerous institutional veto points also tend to have weaker climate policies, as costly climate regulations are more difficult to implement than in countries where power is more concentrated (Lachapelle and Paterson 2013, Madden 2014).

Likewise, the party politics literature identifies a range of characteristics that make political parties more or less likely to embrace climate change as an issue. Right-wing parties tend to have weaker positions on climate change than left-wing parties, as they are typically more averse to the state intervention and market regulation that climate policy often warrants (McCright and Dunlap 2011, Båtstrand 2014, Farstad 2017). Smaller parties have a strategic incentive to emphasise niche positions such as climate change, both to attract voters and to achieve policy differentiation (Wagner 2012, Spoon *et al.* 2014, Abou-Chadi and Orlowski 2016). Lastly, the party competition literature suggests that opposition parties will more likely emphasise new issues such as climate change, as they will be eager to find ways of attacking the government (Klingemann et al. 1995, p. 28, Carter 2006). How do country and party characteristics interact to determine climate policy ambition, and importantly how does country size influence this relationship?

The expectations with regards to the effect of small country size are largely based on Katzenstein's (1985) seminal work Small states in world markets. In examining the successes of small and economically vulnerable nations in western Europe, he found that they had managed to stay competitive and stable through what he terms 'democratic corporatism' – a mixture of bi-partisan consensus, centralised politics and coordination between politicians and businesses. This setup is an effective way of coping with a rapidly changing world, and also makes it possible to shield particularly vulnerable groups or industries through state spending (also see Keating 2015). Significantly, as discussed above, such corporatist coordination has been found to have a positive impact on climate policy. Although he was writing about political economy, the institutional setup outlined by Katzenstein is likely to facilitate cross-party consensus on the issue of climate change as well, through incentivising and making it easier for parties to cooperate in the face of such a global challenge. As Katzenstein (2003, p. 11) articulates it – 'perceived vulnerability' generates an 'ideology of social partnership' that acts 'like a glue' for the corporatist politics of the small European states. Keating (2015, p. 16–17) also underlines the benefits of short lines of com-munication and shared values and trust amongst the governing elite in small states. Moreover, small states often see the environment and climate change as ways of gaining soft power and improving their reputation through international policies and diplomacy (Ingebritsen 2002). Lastly, Katzenstein (2003, p. 19) outlines how small European states build high levels of political confidence, stability and trust by shielding vulnerable groups from interna-tional problems through state spending and social protection. This might also aid the building of consensus on climate change, as voters and parties will presumably be less hostile to potentially damaging climate policies when such a relationship of social protection and trust exists. This relationship can also be conjectured from similar research which shows that economically 'kinder, gentler societies' perform better in protecting the environment (Rootes *et al.* 2012, Bernauer and Böhmelt 2013).

However, Germany is a large country with corporatist institutions, and also experiences cross-party consensus on its ambitious climate polices. This case should make us question the 'size matters' hypothesis and ask if the overriding factor shaping a country's climate politics is not simply their institutional makeup, independent of size.

On the other hand, corporatism is not essential for the creation of cross-party consensus on climate change, as the large state case of the UK demonstrates. Only three Members of Parliament (MPs) voted against the UK's groundbreaking Climate Change Act. Moreover, the effects of corpor-atism are not necessarily positive, with Keating and Harvey (2014, p. 59) highlighting that economic coordination may in some cases impede sig-nificant reform and neglect larger problems. As small state economies are less diverse and can be more reliant on certain sectors than in larger states, corporatism can sometimes lead to deals being made between a particular industry and the government, and protection of the status quo (Lehmbruch 1984). Furthermore, the smallness of countries may actually limit their climate ambitions if parties and voters believe action to be futile due to their limited contribution to global greenhouse gasses (Thorhallsson and Wivel 2006, p. 654). Katzenstein's hypotheses about the benefits of democratic corporatism in the face of economic vulnerability do not necessarily extend to climate vulnerability, as the incentives for responding to the two problems will be quite different. Small states have a strong incentive to respond to the former, but not necessarily the latter given its less immediate and more long-term impacts.

A key question is therefore whether the beneficial outcomes for party agreement and ambitious climate policy stem from the smallness of these countries per se or from their institutional makeup? It is worth unpacking whether it is the size of a state that matters for the party politics of climate change, or whether certain institutional features are simply held in common between small and large states. The case studies used to explore this question are outlined in the next section.

Outline and justification of the cases

Using the logic of the 'controlled comparison' and the 'Method of Difference', the cases of Australia and Norway are selected for analysis. The two countries experience different levels of party agreement and ambition on climate change policy yet share several significant commonalities that provide a fruitful basis for comparison (indeed, these countries have been successfully compared elsewhere in the comparative climate policy literature – see Eckersley 2013). Using this method thus allows us to exclude similar characteristics from the analysis and focus on the differing factors, which should necessarily form part of the explanation for variation.

Australia's emissions reduction targets range in the lower band of developed countries, despite Australia being one of the highest emitters on a per-capita basis in the OECD (IEA 2012, p. 33). Climate change denialism is not uncommon in the right-wing Liberal and National parties (Talberg and Howes 2010, Fielding *et al.* 2012), which has led to intense disagreements along major party lines over 'the urgency of the problem, Australia's international responsibility vis-à-vis developing countries, the type and degree of engagement with the multilateral climate negotiations and the choice of climate policy instruments' (Eckersley 2013, p. 390). Although a Labor government, with support from Greens and independent MPs, in 2011 introduced a comprehensive

Carbon Pricing Mechanism as an intended precursor to an emissions trading scheme, Australia famously became the first country in the world to dismantle a functioning carbon market, with then Prime Minister Tony Abbott labelling the so-called 'carbon tax' 'a great big tax on everything' and 'a so-called market in the non-delivery of an invisible substance to no one' (Sydney Morning Herald, 15 July 2013). With this mind, Tranter argues that party polarisation on climate change constitutes 'one of the strongest impediments to progressive climate change policy' in Australia (2013, p. 411).

In contrast, Norway has set itself the ambitious target of at least 40% emissions reductions below 1990-levels by 2030 under the Paris Agreement, with the aim to be carbon neutral by 2050. The petroleum industry faces strict environmental regulations, and a carbon tax was introduced as early as 1991 (the price doubled in 2012). An emissions trading scheme (ETS) was introduced in 2005, and was incorporated into the EU ETS in 2008. Significantly, Norway's ambitious climate goals are supported in the Storting by six out of seven parties (the exception being the Progress Party). The cross-party climate settlement ('Klimaforliket') reached in 2008 was strengthened in 2012, creating stability and predictability around climate change policy.

Despite the variation in climate politics, the countries share important similarities, which provide a fruitful basis for comparison. Both countries are sparsely populated, developed and wealthy democracies, with similar standards of living (UNDP 2015) and also have a similar quality of democracy (Polity IV 2016). Both countries are highly integrated into world society and the global economy, yet neither country was significantly affected by the global financial crisis (GFC) and both have relatively low levels of unemployment (ABS 2016, SSB 2016).[1] Significantly, Australia and Norway are both major fossil fuel exporters. Australia is the world's largest coal exporter (International Energy Agency (IEA) 2012) and Norway is the third largest exporter of energy in the world with the petroleum sector constituting 47% of exports (IEA 2011, p. 13). Moreover, although the countries' domestic energy profiles differ substantially – Australia's electricity generation is still heavily dependent on coal whereas Norway largely relies on hydro-electricity – they nonetheless face similarly high marginal abatement costs for emissions reductions. As Norway's domestic energy production is already essentially decarbonised, it can only reach its emissions reduction targets by reducing emissions from the petroleum, manufacturing and transport sectors, which already operate at high levels of efficiency (IEA 2011, p. 7–9). Thus Australia and Norway share significant commonalities that are consequently controlled for when seeking to explain the variation in climate politics.

The countries also differ in significant ways. Australia is large (7.69 million km^2) whilst Norway is smaller (385,252 km^2). Australia has a federal political system with multiple veto points whereas Norway is a unitary state with few institutional veto points (Political Constraint Index, NSD 2011). Australia is pluralist (Marsh 2007) whilst Norway is corporatist (Dryzek *et al.* 2002). Lastly, Australia effectively has a mainly majoritarian two-party system (Reynolds *et al.* 2005) whilst Norway has a multi-party proportional electoral system (Arter 2008). These differing characteristics are not controlled for and are thus likely to form part of the explanation of why climate politics differs between the countries.

Importantly, the two cases should help illuminate other, similar, cases – with the findings from Norway expected to be generalisable to other small or corporatist western European states (such as the Scandinavian countries or Germany), and the findings from Australia hopefully shedding light on other large developed and fossil fuel producing states (such as Canada and the USA).

The party politics of climate change: size and institutions

To explore the reasons for the differing party politics of climate change in both countries – and, importantly, to unpack the effects of state size and institutions – 44 semi-structured interviews were conducted with politicians, ENGO and fossil fuel industry representatives, civil servants and policy advisors, and academics/experts in both countries. Twenty-two interviews were conducted in Australia in 2015 and 22 interviews were conducted in Norway in 2016 (Appendix). A qualitative approach has several benefits in this case. First, given the limited number of countries to compare and the lack of cross-national and comparable data, an in-depth and qualitative investigation is preferable. Second, given how inextricably linked state features are – and the remaining amount of differing country characteristics outlined in Table 1 – a qualitative approach is helpful in separating the effect of each factor. A qualitative analysis allows us to identify where and why the 'size matters' hypothesis stands up, and where and why other factors have a stronger influence, thus allowing the researcher to differentiate between the effects of state size and institutions more confidently than through a quantitative analysis. Actors were selected on the basis of their expert insight into the policy area and asked questions about: their perceptions of public opinion; on the role of the fossil fuel industry and interest groups; the dynamics of party competition and country size; and their views on national and international climate policy. Based on the interviews, key themes emerged. The supporting evidence is shown in the parentheses, indicating the number of actors who highlighted a particular theme, and often exemplified with relevant quotes.

Table 1. Comparability of the cases.

Country characteristics	Australia	Norway
Population density	Low	Low
GDP pc	High	High
Standard of living	High	High
Quality of democracy	High	High
International integration	High	High
Effects of GFC	Weak	Weak
Unemployment levels	Low	Low
Fossil fuel exportation	High	High
Marginal abatement costs	High	High
Size	Large	Small
Number of veto points	High	Low
Interest aggregation	Pluralism	Corporatism
Electoral system	Majoritarian	Proportional
Party politics of climate change	Polarised/weak	Consensual/ambitious

Size matters?

As both countries have large and regionally concentrated fossil fuel industries, we would expect this to influence the propensity of politicians in both countries to embrace climate policies that might adversely impact such interests. Furthermore, in line with the expectations of the party politics literature, we would expect this to particularly influence regulation-averse right-wing politicians. However, the interviews conducted in Australia and Norway indicate that state size helps moderate this relationship.

The Australian interviewees highlighted the strong regional focus to public attitudes on climate change, with marked differences between groups and regions (Interviews A1, A3, A4, A5, A9, A13, A16, A18). According to the interviewees, this regional variation incentivises right-wing parties in particular to act as veto points for the policies that would adversely affect their state or constituent interests (Interviews A1, A3, A4, A7, A9, A12, A15, A20). As an example, a Liberal MP claimed that protecting constituent interests was paramount: 'I mean that's your job! You must represent your constituency. You know, you have a view for Australia and the broader interest, but your primary job is trying to deliver a better world for your constituents' (Interview A3). A former Labor cabinet minister argued that politicians seeking to protect regionally-based fossil fuel interests were 'absolutely a large part' of the reason why Australia was so polarised on the issue (Interview A7). Similarly, an ENGO representative argued that states such as Tasmania, Queensland and Western Australia acted as 'huge barriers' to the creation of climate policy and consensus due to the strength of mining and forestry interests there (Interview A12). Likewise, a Labor MP argued that: 'state governments in particular are into "boosterism" – they like to promote economic activity in their state, almost no matter what it is' (Interview A6). This was echoed by a senior civil servant: 'In every field of policy, states are looking out for themselves, and the Commonwealth has really very little except the taxation power to deal with them to try to persuade them to come together and do things in a harmonious way' (Interview A20). Similarly, a former cabinet minister from the conservative National Party pointed out that the only input state branches of his party had on federal climate policy was: 'Don't you hurt our economy, don't you shut our industries down!' (Interview A1). Naturally the binary majoritarian electoral system in Australia influences this relationship, with MPs mindful of the potential of being ousted by political rivals if they are seen to harm constituent interests. However, the sheer geographical size of Australia underpins the variation in constituent interests, and thus seems to play a part in fuelling party polarisation on climate policies by making it harder for right-wing politicians to prioritise national interests over regional or constituency ones.

We would expect Norway, like Australia, to be characterised by regional variation in public attitudes towards climate change. The western regions around Bergen and Stavanger are heavily dependent on the fossil fuel sector for employment (and indeed Tvinnereim and Austgulen (2014) find that Norwegians working with fossil fuels show significantly less acceptance of mainstream climate scientific findings than the general population), and being a similarly sparsely populated country one would expect rural voters to be averse to climate policies that disproportionately affect them (e.g. transport policies regulating the cars on which they depend) (see Norgaard 2011). However, reflecting Keating's (2015, p. 16–17) argument regarding the benefits of short lines of communication in small states, the Norwegian interviews showed that the absence of any laggard or veto region was partly due to the close relationship between the party branches across levels of government. Whereas the Australian politicians interviewed admitted to not consulting state branches or local members on climate policy and emphasised their differences of opinion, Norwegian politicians emphasised: the close links between the national and local branches; how important local politicians were in the development and formulation of national climate policy; and how the various branches of the party were in agreement on the issue (Interviews N3, N4, N5, N6, N7, N8, N9, N10, N11, N12, N13, N14). As one Labour MP expressed it: 'I feel we speak as one team when it comes to climate change (…) We're a movement, so our policies are created from the interaction of the central and the local (…) and we largely use our local politicians in the process of policymaking' (Interview N10). Even the agrarian Centre Party, which places the protection of rural interests at its ideological core (e.g. 'I think we more than any party in the Storting will be concerned that what we legislate doesn't have a negative impact on rural areas' – Interview N8), admitted to maintaining strong party lines on climate change despite some disagreements: 'In any party there'll be disagreements on how we balance climate change with the economy and employment (…) and we have debates and votes at national conferences where issues are decided by a very tiny majority in the end, but then that decision becomes quite determining for the parliamentary group' (Interview N7). Thus the small size of Norway seems to improve the lines of communication between party branches across levels of government, thereby more efficiently overcoming interregional differences of interests and values on climate change, and making it easier to create consensus on ambitious policies.

Institutions matter?

Though the moderating effect of state size on the parties' behaviour permeated the interviews in each country, the impact of the countries' institutional makeup was even more apparent – and in particular the effects of their pluralist and corporatist features.

Underlining a common feature of pluralist systems, the Australian interviews highlighted the strong and privileged position of the fossil fuel lobby in comparison to ENGOs (Interviews A4, A7, A10, A11, A12, A13, A14, A15, A16, A17, A20, A21). A member of the Climate Change Authority (which provides independent advice to government) argued that although most Australian fossil fuel companies have shifted their position on climate change due to the growing momentum globally to tackle the issue, they nonetheless only 'half go along with it' in public and 'behind the scenes still engage in blanket denial and resistance' (Interview A21). Similarly, a former Labor cabinet minister outlined how opposition from emissions-intensive businesses had been the 'main challenge' for the government when developing its climate policies: 'There were massive challenges, and outright hostility and opposition from many sectors of the business community, particularly the fossil fuel industries' (Interview A7). Importantly, the hostility from certain fossil fuel companies towards more stringent regulations or targets is combined with easy access to and power over decision-makers (Interviews A10, A12, A13, A15, A16, A17, A20, A21). The member of the Climate Change Authority argued that the fossil fuel lobby's access is 'very strong', and called them 'the most powerful lobby in Australia.' 'They're very, very effective. They spend a lot of money and they get the best lobbyists' (Interview A21). Even the representative of the Australian Industry Greenhouse Network (an industry association representing large emitters) admitted that the fossil fuel industry, and especially the Minerals Council, engages in 'really extensive campaigns with lots of politicians' (Interview A16). Furthermore, a representative from the Construction, Forestry, Mining and Energy Union argued that political parties in Australia develop their stances and policies on climate change 'based on how they are lobbied' – 'basically the views given to them by lobby groups and vested interests' (Interview A17). A Green Party politician outlined a similar relationship:

> They've [the fossil fuel lobby] got a lot of money, and they can afford to donate to political parties. And they can threaten to unseat members of parliament – or in some instances, as we found when we were debating the mining tax, prime ministers – if they don't do what they want. So as a result, many members of parliament live in fear of a campaign being run to unseat them. And, conversely, they're quite happy to receive donations from these companies. So one senator got up during a debate in parliament proudly wearing a vest that said 'Australians for coal' which the coal lobby had prepared (...) for those members of parliament that they felt had advocated for them or were, some might say, 'in their pocket'. So it's the equivalent of a football player wearing their sponsor's logo. So in many respects it's that blatant. It's bankrolling and the threat of punishing (Interview A10).

The power of the fossil fuel lobby stands in stark contrast to that of ENGOs (Interviews A11, A12, A13, A15, A20, A21). A central concern of all the ENGOs interviewed was limited funding and the impact this had on the effectiveness of their lobbying. A representative from Friends of the Earth (FoE) Australia pointed out that ENGOs needed resources in the capital, Canberra, in order to be effective at the federal level, but 'many of the green groups don't have those.' With a lack of such resources FoE had 'focused downwards on the state level' which put them at a comparative disadvantage to the fossil fuel lobby (Interview A13). Likewise, a representative from an influential ENGO said 'we just put out a story to the media last week basically saying "help, we're running out of money!"' (Interview A15). A representative from The Wilderness Society claimed the 'hostile' and 'deeply anti-conservationist, pro-business, pro-coal, pro-fossil fuel' Abbott government had tried to remove the tax-deductibility status of ENGOs in a bid to limit their funding and debilitate them (Interview A12), and the FoE representative, agreeing, claimed this was done 'at the behest of the mining industry' (Interview A13). The picture that is painted is thus one of limited funds and effectiveness of ENGOs relative to the fossil fuel industry, highlighting a common characteristic of pluralist systems.

The hostility of the fossil fuel industry towards more stringent climate policies could perhaps have been avoided by a more corporatist approach, as in the Norwegian case. The Norwegian petroleum industry is considerably more constructive when it comes to climate policy than its Australian counterpart, making the creation of party agreement easier to achieve. Although the industry shows no signs of foregoing the benefits of production or expansion, it is nonetheless supportive of strict regulations and ambitious emissions targets. According to the interviewees, this constructive attitude has developed as a result of the close relationship between state and industry (Interviews N3, N4, N5, N6, N7, N8, N9, N10, N11, N12, N13, N14, N15, N19, N20, N21, N22).

However, the interviews revealed that it is not simply the relationship between state and industry that creates a constructive dialogue on climate change but, importantly, the close relationship between the state and ENGOs as well. In contrast to Australia, Norwegian ENGOs receive state funding and thus have more resources at their disposal. Their high levels of activity and effectiveness at lobbying were noted by several actors (Interviews N3, N4, N5, N6, N7, N8, N9, N10, N11, N12, N13, N14, N15, N16, N17, N18, N19, N21). For example: 'What surprised me the most when I became an MP was that I was expecting the business and fossil fuel lobby to be very strong, but that isn't correct at all. The people who come here are the ENGOs – they're the ones who lobby the most and are the most professional. They have a huge apparatus. (…) They're here all the time and we get weekly requests for meetings' (Interview N4). Similarly, several interviewees pointed out how ENGOs were more active at committee hearings, and were almost overrepresented (Interviews N3, N4, N15, N16, N17, N18, N19, N22).

Not only is the ENGO lobby strong, but its relationship to and dialogue with the business community are also better than in Australia. A wide range of stakeholders interacts regularly in an institutionalised setting, for example on the Minister of Climate and Environment's Climate Council (Klimarådet). Several interviewees highlighted the frequency of interactions and the quality of the dialogue with 'opposing' stakeholders, as well as the benefits this had (Interviews N15, N16, N17, N18, N19, N20). As one member of the Climate Council outlined:

> It's great when (…) there's a recognition around the table that "we're actually working towards a common goal even if we have different standpoints, tools and priorities." But the key thing that happens is that a deeper understanding develops between us all, especially between businesses and ENGOs, or between the private and public sector, or between academia and businesses. So there's a deeper understanding and a deeper recognition, and that creates the basis for a better dialogue, and that's good (Interview N20).

Other members of the Climate Council echoed this sentiment. A WWF representative argued: 'Having forums like this where issues are brought up and suggestions and knowledge are sought is nice, but it's especially useful to notice what the other groups are saying' (Interview N15). Similarly, a Norwegian Confederation of Trade Unions (LO) representative pointed out: 'ENGOs now get the need for a just transition and that it's not black-or-white, or for-and-against. So the more we talk to each other the more we understand each other, and the more holistically we can see these things, the easier it is to get climate policy through' (Interview N19).

Lastly, dialogue between the state and opposing stakeholders was not the only corporatist feature highlighted as being beneficial for the creation of consensus on climate policy. One characteristic highlighted by Katzenstein (1985), the state's willingness to shield or compensate losers, also helps create agreement. In this respect, Norwegian welfare arrangements were critical in improving the constructiveness of businesses. As the LO representative pointed out:

> If you're in danger of losing your job then the resistance will be so big that you can't make any changes. If you talk to miners in Australia, the US or Poland, you'll know what I mean. If there's no alternative to your job then you're obviously going to cling to the status quo. Because there's nothing else to go to, the green transition won't happen. That's why we use the term 'just transition', because if mass unemployment is the alternative then you're not going to be able to make the necessary changes. But it's been easier here because we have good welfare provisions and safety nets (at least for the time being), which mean that if you lose your job you won't end up on the street or having to sell your house. So if you have those kinds of tools you can create and implement a green transition in a completely different way (Interview N19).

Thus with substantial welfare provisions there is less opposition from exposed industries, reducing their hostility to climate policies and improving the prospects for dialogue, compromise and cross-party consensus.

Discussion

Does state size influence the party politics of climate change, and does Katzenstein's work help us understand the differences between the climate politics of Australia and Norway?

The Australian interviews indicate that the large size of the country makes it harder to overcome interregional differences of interests and values on climate change – both as a consequence of a lack of feeling of national unity and right-wing politicians wanting to prioritise regional and constituent interests, and because of longer and poorer lines of communication. The benefits of shorter lines of communication are highlighted in the Norwegian case. With closer proximity and increased interaction between levels of government, regional fossil fuel interests have posed less of an obstacle to the creation of cross-party consensus.

However, institutional factors are found to play a more significant role in explaining the variation between the countries. The interviews demonstrate that the pluralist features of Australian politics award the fossil fuel industry a privileged position in terms of both resources and access to policymakers in comparison to ENGOs. This fails to check or counterbalance the industry's hostility to more stringent regulations and targets, and awards them more influence on regulation-averse right-wing politicians. In Norway, the corporatist features of institutionalised cooperation between state and industry as well as influential green groups have made the fossil fuel industry more constructive on climate change. Combined with generous welfare provisions, these features have facilitated the creation of cross-party consensus on climate change.

As such, the institutional setup outlined by Katzenstein – 'democratic corporatism' – does indeed help us understand why smaller Norway has more ambitious climate policy than Australia. However, state-business cooperation and trust-inducing welfare arrangements, although common features of small states, are not directly related to size per se, and only the short lines of communication in Norway can be directly linked to the size of the state. However, as mentioned, the benefits of small state size permeated the Norwegian interviews far less than the benefits of their corporatist features. Furthermore, cross-party cooperation in the Norwegian case was to a large extent incentivised by the parties' exposure to a range of lobby groups as a consequence of this institutional setup, and a perceived vulnerability of the country to external forces was not mentioned, contrary to what Katzenstein (1985, 2003)) would predict. In the Australian case, it remains unclear whether right-wing politicians protect regionally-based fossil fuel interests primarily due to the size of the country and feelings of loyalty to constituents, or because of the influence of the fossil fuel lobby over them as a result of the pluralist system of government – though the interviews point towards the latter explanation. Thus as institutional features seem to be influencing the parties' behaviour more than state size per se, the 'size matters' hypothesis – although not disproven – finds somewhat limited support in the two case studies. Interestingly, to the limited extent that state size does seem to matter, it is the geographical – not economic – size of the country that seems to be shaping the behaviour of the parties, contrary to what Katzenstein's work would lead us to hypothesise.

However, despite the in-depth nature of the current analysis, given how inextricably linked state features are, it remains difficult to separate the effect of each factor, and further research is required to test the generalisability of the findings and differentiate the effects of state size and various institutions with more confidence. One useful avenue of future investigation would therefore be to compare further cases. Such case studies should also focus on different versions of corporatism and pluralism, to explore whether certain features therein are more important than others. The contributions to the present volume are one step in this direction. A second avenue of future investigation would be to complement qualitative studies with larger cross-national quantitative analyses, in order to pinpoint the effects of each factor. This could either be done through regression analyses (data availability allowing), or through medium-N qualitative comparative analyses (QCA) to examine whether individual factors are needed separately or in conjunction to influence the outcome (Ragin 2000).

Such research would also help fill gaps in the current investigation. Although the countries have been compared fruitfully, the 'controlled comparison' of Australia and Norway has limitations. Led by the inter-views themselves – and importantly as these features are critical in testing the applicability of Katzensetin's work and the 'size matters' hypothesis – the current investigation has focused on comparing the effects of state size and of pluralist and corporatist features. However, these are only two of the differing characteristics between the countries highlighted in Table 1, and other differing characteristics (e.g. the number of veto points and the electoral system) could be confounding variables explaining the relationships exposed in both countries. For example, are strong party lines in Norway maintained as a result of short lines of communication or as a result of few veto points? Similarly, do the Norwegian parties converge as a result of their cor-poratist features, or is it the impact of the proportional electoral system increasing the incentives to compete and compromise on climate change? Or both? Although the benefit of in-depth, qualitative research is to differentiate such effects – and to a large extent the current analysis does so in that these effects do not feature to the same extent as those highlighted here – their confounding influence cannot be excluded.

Finally, a significant contribution of the current investigation is the explora-tion of the effects of corporatist and pluralist features for party agreement on climate change, and the identification of their critical role in shaping the parties' behaviour. The outlined benefits of corporatism and detrimental effects of pluralism support previous work in the comparative climate policy literature, with the exploration at the party-level being novel. The party-level analysis thus also feeds into the growing literature on the party politics of climate change. In addition, I have provided an important contribution to the growing literature seeking to understand the reasons behind party polarisation on climate change and the environment. My findings should therefore help shed light on similar relationships in other small and large states respectively.

Conclusion

The burgeoning comparative climate policy literature has largely neglected both the role of state size and political parties in explaining variation in countries' climate policies and ambition. These omissions are regrettable, as political parties – and in particular bi-partisan agreement – are critical in achieving stable and ambitious climate policies.

Building on the expectations provided by the comparative climate policy, party politics and state size literature – and in particular Katzenstein's seminal work on small states – I have tested the 'size matters' hypothesis through an in-depth and qualitative examination of two cases, Australia and Norway. Following the logic of the 'controlled comparison' these are fruitful countries to compare – they are similar in important respects yet differ in terms of their size as well as their level of party agreement and ambitions on climate change. In particular, the review of original primary data from 44 interviews with policymakers and policy-shapers in both countries has shed light on the factors that facilitate or impede cross-party consensus on climate change.

I have revealed that, to the limited extent that state size matters, it is the geographical – not economic – size of the state that moderates the parties' behaviour. The large size of Australia makes it harder to overcome interregional differences in interests and values, and thus impedes the creation of cross-party consensus on climate change, whereas the small size of Norway facilitates consensus through short lines of communication. However, the pluralist and corporatist features in Australia and Norway respectively have been shown to have an overriding effect on the party politics of climate change in each country. In particular, the benefits of state-business cooperation, strong ENGOs and compensatory welfare arrangements have been highlighted as beneficial for the building of consensus. These findings add a party-level analysis to the existing comparative climate policy literature on the topic. However, the pluralist and corporatist features examined – although common in large and small states respectively – are not strictly related to state size per se. Thus overall, the 'size matters' hypothesis finds somewhat limited support here. Further research is warranted to unpack the effects of state size and various institutions, and to pinpoint the effects of each with more confidence.

Note

1. For that reason, we can rule out the recession as providing the explanation for polarisation in Australia.

Disclosure statement

No potential conflict of interest was reported by the author.

Funding

The article draws on research conducted as part of an ESRC-funded PhD 'From consensus to polarisation: What explains variation in party agreement on climate change?'

References

Abou-Chadi, T. and Orlowski, M., 2016. Moderate as necessary: the role of electoral competitiveness and party size in explaining parties' policy shifts. *The Journal of Politics*, 78 (3), 868–881. doi:10.1086/685585

Arter, D., 2008. *Scandinavian politics today*. 2nd ed. Manchester: Manchester University Press.

Australian Bureau of Statistics (ABS), 2016. *Labour force* [online]. Available from: http://www.abs.gov.au/ausstats/abs@.nsf/mf/6202.0 [Accessed 18 September 2016].

Båtstrand, S., 2014. Giving content to new politics: from broad hypothesis to empirical analysis using Norwegian manifesto data on climate change. *Party Politics*, 20 (6), 930–939. doi:10.1177/1354068812462923

Båtstrand, S., 2015. more than markets: a comparative study of nine conservative parties on climate change. *Politics and Policy*, 43 (4), 538–561. doi:10.1111/polp.12122

Bernauer, T., and Böhmelt, T., 2013. National climate policies in international comparison: the climate change cooperation index. *Environmental Science And Policy*, 25, 196-206.

Bernhagen, P., 2008. Business and international environmental agreements: domestic sources of participation and compliance by advanced industrialized democracies. *Global Environmental Politics*, 8 (1), 78–110. doi:10.1162/glep.2008.8.1.78

Carter, N., 2006. Party politicization of the environment in Britain. *Party Politics*, 12 (6), 747–767. doi:10.1177/1354068806068599

Carter, N., *et al.*, 2017. Political parties and climate policy: a new approach to measuring parties' climate policy preferences. *Party Politics*. doi:10.1177/1354068817697630

Ćetković, S. and Buzogány, A., 2016. Varieties of capitalism and clean energy transitions in the European Union: when renewable energy hits different economic logics. *Climate Policy*, 16 (5), 642–657. doi:10.1080/14693062.2015.1135778

Dryzek, J.S., *et al.*, 2002. Environmental transformation of the state: the USA, Norway, Germany and the UK. *Policy Studies*, 50 (4), 659–682.

Dunlap, R.E., McCright, A.M., and Yarosh, J.H., 2016. The political divide on climate change: partisan polarization widens in the U.S. *Environment: Science and Policy for Sustainable Development*, 58 (5), 4–23.

Eckersley, R., 2013. Poles apart? The social construction of responsibility for climate change in Australia and Norway. *Australian Journal of Politics and History*, 59 (3), 382–396. doi:10.1111/ajph.12022

Fankhauser, S., Gennaioli, C., and Collins, M., 2015. The political economy of passing climate change legislation: evidence from a survey. *Global Environmental Change*, 35, 52–61. doi:10.1016/j.gloenvcha.2015.08.008

Farstad, F.M., 2017. What explains variation in parties' climate change salience? *Party Politics*. doi:10.1177/1354068817693473

Fielding, K.S., *et al.*, 2012. Australian politicians' beliefs about climate change: political partisanship and political ideology. *Environmental Politics*, 21 (5), 712–733. doi:10.1080/09644016.2012.698887

Griffiths, A.B., Haigh, N., and Rassisas, J., 2007. A framework for understanding institutional governance systems and climate change: the case of Australia. *European Management Journal*, 25 (6), 415–442. doi:10.1016/j.emj.2007.08.001

Harrison, K. and Sundstrom, L.M., eds., 2010. *Global commons domestic decisions: the comparative politics of climate change*. Cambridge, MA: MIT Press.

Ingebritsen, C., 2002. Norm entrepreneurs: Scandinavia's role in world politics. *Cooperation and Conflict*, 37 (1), 11–23. doi:10.1177/0010836702037001689

International Energy Agency (IEA), 2011. *Energy policies of IEA countries: norway* [online]. Available from: http://www.iea.org/publications/freepublications/publication/energy-policies- of-iea-countries—norway-2011-review.html [Accessed 16 September 2016].

International Energy Agency (IEA), 2012. *Energy policies of IEA countries: Australia* [online]. Available from: http://www.iea.org/publications/freepublications/publication/energy-policies-of-iea-countries—australia-2012-review.html [Accessed 21 September 2016].

Ireland, J., 2013. Emissions scheme a trade in the invisible: Abbott. *Sydney Morning Herald*, 15 July.

Jensen, C.B. and Spoon, J.J., 2011. Testing the 'party matters' thesis: explaining progress towards Kyoto protocol targets. *Political Studies*, 59, 99–115. doi:10.1111/j.1467-9248.2010.00852.x

Katzenstein, P.J., 1985. *Small states in world markets: industrial policy in Europe*. New York: Cornell University Press.

Katzenstein, P.J., 2003. Small states and small states revisited. *New Political Economy*, 8 (1), 9–30. doi:10.1080/1356346032000078705

Keating, M., 2015. The political economy of small states in Europe. *In*: H. Baldersheim and M. Keating, eds. *Small states in the modern world : vulnerabilities and Opportunities*. Cheltenham, Gloucestershire: Edward Elgar, 1–19.

Keating, M. and Harvey, M., 2014. The political economy of small European states: and lessons for Scotland. *National Institute Economic Review*, 227 (1), R54–R66.

Klingemann, H.D., Hofferbert, R., and Budge, I., 1995. *Parties, policies and democracy*. Boulder, CO: Westview.

Lachapelle, E. and Paterson, M., 2013. Drivers of national climate policy. *Climate Policy*, 13 (5), 547–571. doi:10.1080/14693062.2013.811333

Lehmbruch, G., 1984. Concertation and the structure of corporatist networks. *In*: J. H. Goldthorpe, ed. *Order and conflict in contemporary capitalism*. Oxford: Clarendon Press, 60–80.

Liefferink, D., *et al.*, 2009. Leaders and laggards in environmental policy: a quantitative analysis of domestic policy outputs. *Journal of European Public Policy*, 16 (5), 677–700. doi:10.1080/13501760902983283

Madden, N.J., 2014. Green means stop: veto players and their impact on climate change policy outputs. *Environmental Politics*, 23 (4), 570–589. doi:10.1080/09644016.2014.884301

Marsh, I., 2007. Interest groups. *In*: B. Galligan and W. Roberts, eds. *Oxford companion to Australian politics*. Oxford: Oxford University Press.

McCright, A.M. and Dunlap, R.E., 2011. The politicization of climate change and polarization in the American public's views of global warming, 2001–2010. *The Sociological Quarterly*, 52, 155–194. doi:10.1111/j.1533-8525.2011.01198.x

Neumayer, E., 2002a. Does trade openness promote multilateral environmental cooperation? *World Economy*, 25 (6), 815–832. doi:10.1111/1467-9701.00464

Neumayer, E., 2002b. Do democracies exhibit stronger international environmental commitment? A cross-country analysis. *Journal of Peace Research*, 39 (2), 139–164. doi:10.1177/0022343302039002001

Norgaard, K.M., 2011. *Living in denial: climate change, emotions and everyday life*. Cambridge, MA: MIT Press.

NSD [Nordic Social Science Data Services], 2011. *Political constraint index dataset* [online]. Available from: http://www.nsd.uib.no/macrodataguide/set.html?id= 29&sub=1 [Accessed 11 March 2015].

Polity IV Project, 2016. *Political regime characteristics and transitions, 1800–2012* [online]. Available from: http://www.systemicpeace.org/polity/polity4.htm [Accessed 18 April 2014].

Ragin, C.C., 2000. *Fuzzy set social science*. London: University of Chicago Press.

Reynolds, A., Reilly, B., and Ellis, A., 2005. *Electoral system design: the new international IDEA handbook*. Stockholm: International IDEA.

Rootes, C., Zito, A., and Barry, J., 2012. Climate change, national politics and grassroots action: an introduction. *Environmental Politics*, 21 (5), 677–690. doi:10.1080/09644016.2012.720098

Scruggs, L.A., 1999. Institutions and environmental performance in seventeen Western democracies. *British Journal of Political Science*, 29 (1), 1–31. doi:10.1017/S0007123499000010

Spoon, J.J., Hobolt, S.B., and De Vries, C.E., 2014. Going green: explaining issue competition on the environment. *European Journal of Political Research*, 53, 363–380. doi:10.1111/1475-6765.12032

SSB (Statistics Norway), 2016. *Labour force survey, seasonally-adjusted figures, June 2016* [online]. Available from: https://www.ssb.no/en/arbeid-og-lonn/statis tikker/akumnd/maaned [Accessed 16 September 2016].

Talberg, A. and Howes, S., 2010. Party divides: expertise in and attitude towards climate change among Australian members of parliament. *CCEP working paper 8.10*. Centre for Climate Economics and Policy, Crawford School of Economics and Government, The Australian National University.

Thorhallsson, B. and Wivel, A., 2006. Small states in the European Union: what do we know and what would we like to know? *Cambridge Review of International Affairs*, 19 (4), 651–668. doi:10.1080/09557570601003502

Tranter, B., 2013. The great divide: political candidate and voter polarisation over global warming in Australia. *Australian Journal of Politics and History*, 59 (3), 397–413.

Tvinnereim, E. and Austgulen, M.H., 2014. Fossil fuel employment and public opinion about climate change. *Stein Rokkan Centre for Social Studies, Working Paper 3-2014*.

UNDP, 2015. *Human development index* [online]. Available from: http://hdr.undp. org/en/content/human-development-index-hdi [Accessed 16 September 2016].

Unruh, G., 2000. Understanding carbon lock-in. *Energy Policy*, 28, 817–830. doi:10.1016/S0301-4215(00)00070-7

Wagner, M., 2012. When do parties emphasise extreme positions?: how strategic incentives for policy differentiation influences issue importance. *European Journal of Political Research*, 51, 64–88. doi:10.1111/j.1475-6765.2011.01989.x

Appendix

Table A1. List of interviewees.

Country		Position	Date
Australia	A1	Previous Cabinet Minister, National Party of Australia	03.11.2015
	A2	Elected Representative, Liberal Party of Australia	03.12.2015
	A3	Elected Representative, Liberal Party of Australia	14.12.2015
	A4	Previous Cabinet Minister, Australian Labor Party	23.11.2015
	A5	Elected Representative, Australian Labor Party	24.11.2015
	A6	Elected Representative, Australian Labor Party	26.11.2015
	A7	Previous Cabinet Minister, Australian Labor Party	26.11.2015
	A8	Elected Representative, Australian Labor Party	27.11.2015
	A9	Elected Representative, Australian Labor Party	05.12.2015
	A10	Elected Representative, Australian Greens	09.11.2015
	A11	Elected Representative, Australian Greens	01.12.2015
	A12	The Wilderness Society	28.10.2015
	A13	Friends of the Earth Australia	25.11.2015
	A14	The Climate Council	02.12.2015
	A15	Representative, Influential ENGO	21.12.2015
	A16	Australian Industry Greenhouse Network	05.11.2015
	A17	Construction, Forestry, Mining and Energy Union	20.11.2015
	A18	Australia Environment Foundation	23.11.2015
	A19	Minerals Council of Australia	17.12.2015
	A20	Senior Civil Servant (Environment)	30.10.2015
	A21	Member of the Climate Change Authority	16.11.2015
	A22	Senior Civil Servant (Environment)	18.11.2015
Norway	N1	Elected Representative, Progress Party	25.04.2016
	N2	Cabinet Minister, Progress Party	10.05.2016
	N3	Cabinet Minister, Conservative Party	22.04.2016
	N4	Elected Representative, Conservative Party	26.04.2016
	N5	Elected Representative, Conservative Party	28.04.2016
	N6	Elected Representative, Liberal Party	25.04.2016
	N7	Elected Representative, Centre Party	20.04.2016
	N8	Elected Representative, Centre Party	03.05.2016
	N9	Elected Representative, Labour Party	14.04.2016
	N10	Elected Representative, Labour Party	10.05.2016
	N11	Previous Cabinet Minister, Socialist Left Party	04.03.2016
	N12	Previous Cabinet Minister, Socialist Left Party	14.04.2016
	N13	Previous Cabinet Minister, Socialist Left Party	03.05.2016
	N14	Elected Representative, Green Party	12.04.2016
	N15	World Wide Fund for Nature	15.04.2016
	N16	Friends of the Earth Norway	18.04.2016
	N17	Norwegian Oil and Gas Association	22.04.2016
	N18	Norwegian Petroleum Society	04.05.2016
	N19	The Norwegian Confederation of Trade Unions (LO)	12.05.2016
	N20	Government Expert Adviser (Climate Change)	24.01.2016
	N21	Senior Civil Servant (Climate and Environment)	22.04.2016
	N22	Senior Civil Servant (Petroleum and Energy)	02.05.2016

Most relevant role mentioned.

Drivers of political parties' climate policy preferences: lessons from Denmark and Ireland

Robert Ladrech (iD) and Conor Little (iD)

ABSTRACT

Political parties are important actors in domestic climate politics. What drives variation in parties' climate policy preferences? To contribute to a growing literature on the party politics of climate change, we focus on the roles of public opinion, party competition, and parties' traditional policy preferences in shaping parties' climate policy preferences in Denmark and Ireland. In case studies that draw on in-depth interviews with policy practitioners, we show how parties respond to public opinion, accommodate issue-owners, and are powerfully constrained and enabled by their existing preferences. These mechanisms also help to explain different responses on climate policy across the left-right spectrum. Competition between mainstream parties is particularly powerful, but can constrain as much as it enables 'greener' climate policy preferences. While climate change may be a distinctive problem, the party politics of climate change features similar incentives and constraints as other domains.

Parties play an important role in structuring and channelling the politics of climate change (Båtstrand 2014, 2015, Marcinkiewicz and Tosun 2015, Carter *et al.* 2018, Farstad 2018). They influence public attitudes (Brulle *et al.* 2012, Guber 2013, Sohlberg 2016) and government policy outputs and outcomes (Jensen and Spoon 2011, Birchall 2014). Party polarization on climate change raises the stakes of political competition on climate policy; it influences threat perceptions and public behaviour, and it is generally understood to lead to greater climate policy delay and less effective, ambitious, or consistent climate policy (Christoff and Eckersley 2011, pp. 442–443, Farstad 2016, p. 35, Sohlberg 2016; see also Farstad 2019 – this volume).

What drives variation in parties' climate policy preferences? This study examines the party politics of climate change in Denmark and Ireland over the past two decades, focusing on the roles of public opinion on climate change, party competition, and intra-party factors, especially parties' other, pre-existing policy preferences. Further, it investigates their role in producing the relationship between left-right preferences and climate policy preferences that has been observed in several studies (e.g. De Blasio and Sorice 2013, Carter and Ladrech *et al.* 2018, p. 736, Farstad 2018).

Using case study methods, including in-depth interviews with climate policy practitioners, it finds evidence that each of these three factors plays a role in parties' responses to climate change. Low public concern about climate change is viewed as a considerable constraint on the development of parties' climate policy preferences, while increased concern provides opportunities for parties to raise its salience and to take stronger climate policy positions. Electorally successful issue-owners provoke accommodative behaviour from other parties, especially those close to them in political space, while competition between larger parties on climate change seems to be a powerful driver of climate policy preferences, but not necessarily towards 'greener' preferences. Existing party policy preferences on traditional policy issues function as an important constraint on – and sometimes an enabler of – parties' climate policy preferences. They have had a particularly constraining effect on the preferences of right-of-centre parties.

We proceed as follows. First, we clarify some key concepts and we outline theory and existing knowledge in relation to the three causal factors of interest. In the second section, we discuss case selection and other methodological choices, followed by two country studies and an analytical discussion. The study contributes to the growing evidence base concerning the determinants of party preferences on climate change and its findings regarding the roles of traditional party preferences may have implications for other 'new' issues. It highlights that, notwithstanding climate policy's distinctiveness, the drivers of party preferences on climate policy have much in common with party politics in other domains.

Drivers of parties' climate policy preferences

A climate change mitigation policy is 'a human intervention to reduce the sources or enhance the sinks of greenhouse gases' (IPCC 2014, p. 4); it plausibly encompasses 'anti-climate' policies too, which increase greenhouse gas (GHG) emissions or reduce GHG sinks (Compston and Bailey 2013). Climate policies include carbon pricing, framework climate legislation, and national strategies, amongst many others.

Climate policy is a matter of environmental protection; in this respect, environmental policy is its 'parent' issue. However, not all environmental policies protect the climate: closing nuclear power stations, for example, protects aspects of the environment, but leads to increased GHG emissions in many contexts (Båtstrand 2014, p.933, Carter and Ladrech *et al.* 2018, p. 734). Climate policies also encompass a wider set of subdomains than typical environmental policies. Thus, climate policy is both related to and distinct from environmental policy; this is also reflected in a recent comparison of various measures of parties' environmental and climate policy preferences (Carter and Ladrech *et al.* 2018, pp.737–739).

Parties' preferences vary, both in the emphasis they place on the climate policy (*salience*) and the *position(s)* that they adopt. In some contexts, increased salience goes hand-in-hand with the development of 'greener' positions on climate policy, but this is not always the case (cf. Carter and Jacobs 2014, Marcinkiewicz and Tosun 2015). Indeed, climate policy is often a positional, partisan issue (Farstad 2018, p. 705); this further distinguishes it from environmental policy, which is often a valence issue. Differences in parties' positions contribute to polarization on climate policy[1]; parties also contribute to variation in the systemic salience of climate policy.

What drives variation in parties' climate policy preferences and, thus, system-level structures of climate politics? There is considerable evidence that climate policy preferences are associated with traditional left-right policy preferences on economic and social issues (De Blasio and Sorice 2013, Carter and Ladrech *et al.* 2018, p.736, Farstad 2018), albeit with some exceptions (e.g. Poland: Marcinkiewicz and Tosun 2015). This association is also present in the case of environmental policy (Rohrschneider 1993, Dalton 2009, Jensen and Spoon 2011, Tosun 2011, Carter 2013). However, further research is required to specify the mechanisms underpinning the relationship between climate policy and left-right preferences (Farstad 2016, pp.249–250). We propose that there are at least three routes through which these mechanisms might operate, and that these mechanisms also offer general explanations for the development of parties' climate policy preferences.

First, parties seek votes and therefore *public opinion* can create an environment that is conducive to developing certain policy preferences (e.g. Klüver and Sagarzazu 2016). Carter and Jacobs' (2014) account of British climate politics in the 2000s highlights the significance of increased public concern about climate change for the development of the main parties' climate policy preferences. Spoon *et al.* (2014) find that parties emphasize environmental issues when public concern about those issues is high. Other suggestive evidence comes from the study of environmental policy in Denmark, where public concern about the environment has led to changes in government policy (Seeberg 2016).

However, public opinion on climate change is unevenly distributed among voters: right-of-centre ideals are negatively associated with belief in human-induced climate change and climate policy support, while left-of-centre ideals are positively associated with these dispositions (Hornsey *et al.* 2016). Moreover, larger, mainstream parties may be better able to respond to shifting voter concerns (Wagner and Meyer 2014, Klüver and Sagarzazu 2016, p. 396). However, our knowledge is limited about the role of public opinion in shaping parties' climate policy preferences (Marcinkiewicz and Tosun 2015, p. 201) and in particular about the perceptions and motivations of party elites that produce them.

Second, in developing their climate policy preferences, parties must take into account the behaviour of other parties. Issue-owners are a potential source of *competition*, but findings on their role are contradictory. Spoon *et al.* (2014) find that the success of Green parties forces other parties, especially those that are ideologically proximate to them, to accommodate their environmental policies. However, Abou-Chadi (2016) finds that stronger Green parties deter other parties from raising the salience of environmental policy. In either case – and as indicated by Spoon *et al.*'s (2014) findings – it is likely that competition from parties with strong preferences on climate change creates unequal incentives for parties of the left and right: Green parties are with few exceptions from the left-of-centre (Carter 2013), while some of the parties that have been most sceptical of climate science and policies are of the far right (Gemenis *et al.* 2012, Båtstrand 2014).

Competition on climate policy may have a particularly strong effect when it comes from mainstream parties, although the evidence-base for this assumption is limited. The paradigmatic case of mainstream competition on climate policy is the UK's brief 'competitive consensus' in the mid-late 2000s, which illustrated the particular importance for a centre-left party of not being outflanked by its centre-right rival (Carter and Jacobs 2014).

Third, party preferences are proximately the product of *intra-party* factors and therefore we can expect preferences on a relatively new issue such as climate policy to be influenced by preferences on traditional issues (Meyer 2013). Broader preferences – on issues such as state intervention (e.g. public ownership, regulation, taxation), collectivism, changes to the policy status quo, and free markets – are often assumed to play a central role in explaining the relationship between left-right preferences and climate policy preferences (Båtstrand 2014, 2015, pp.540–542, Farstad 2018). This assumption is supported by some limited empirical evidence on left-right differences from Norwegian manifestos in 2009 (Båtstrand 2014) and from nine conservative parties, whose preferences are constrained by their support for fossil fuel-producing industries (Båtstrand 2015). However, intra-party politics is arguably among the least-well understood factors that shape parties' climate policy preferences.

Case selection and data

To examine the role of public opinion, party competition, and existing policy preferences in parties' climate policy responses, including their role in the relationship between left-right and climate policy preferences, we focus on Denmark and Ireland (see Andersen and Nielsen 2016, Little and Torney 2017 for reviews). We gather data on the period from the mid-1990s to 2016. This is long enough to observe individual parties over several election cycles, governmental configurations, and a variety of economic and international conditions.

During this period, Ireland's climate politics has been characterised by low salience and broad consensus (Little 2017a), while Denmark's has been characterised by greater salience and polarization. Denmark has also seen fluctuations in polarization driven by abrupt changes in climate policy preferences on the centre-right (Eikeland and Inderberg 2016, Seeberg 2016). Applying Dalton's (2008, p. 906) Polarization Index (PI)[2] to parties' positions on two expert-coded climate policy items from 2009 and 2014[3] supports the observation that parties are more polarized on climate policy in Denmark (PI = 3.7 and 3.4 in 2009 and 2014, respectively) than in Ireland (2.9 and 3.2). It is also reflected over a longer period in Carter *et al.*'s (2018) manifesto-based data on climate policy positions for the two main parties in each country: in Denmark, the mean gap between the two main parties' positions from 1994 to 2015 was 2.3 times as large as in Ireland from 1997 to 2011.

Denmark and Ireland's general similarities have provided a basis for several comparative case studies of economic policies (e.g. Giavazzi and Pagano 1990, Kluth and Lynggaard 2013, Campbell and Hall 2015). They are also similar in ways that bear on parties' incentives and constraints in developing climate policy preferences. They are long-established parliamentary democracies with proportional representation electoral systems and few veto points; these systems and the parties in them are most likely to respond to the challenge of climate change mitigation (Christoff and Eckersley 2011, p.440, Lachapelle and Paterson 2013, p.549, Spoon *et al.* 2014). Both have been EU member states since 1973, accounting for an important set of supranational constraints and incentives for domestic climate policy actors.

In terms of problem-pressure, both countries have high per capita emissions but relatively low vulnerability to climate change impacts (e.g. Eckstein *et al.* 2017). They have open economies and both have large, export-oriented agriculture sectors, which account for significant proportions of their GHG emissions: typically more than 30% in Ireland and approximately 20% in Denmark. Nuclear energy has been off their respective political agendas for some time and both experienced a property market and banking crisis in the late 2000s, albeit only Ireland received a multilateral 'bailout' loan (Kluth and Lynggaard 2013, Campbell and Hall 2015).

Despite their similarities, Denmark and Ireland differ in some important respects for the purposes of this study. First, Denmark's party system is centred primarily on the left-right dimension, while Ireland's party system is structured by two centrist parties in which both the left and the far right are relatively underdeveloped. Applied to expert survey and manifesto data on left-right positions (Bakker *et al.* 2015, Volkens *et al.* 2017), Dalton's (2008) PI supports these observations, with consistently greater left-right polarization evident in Denmark than Ireland.[4] This has implications for both party competition and for the intra-party politics of climate policy in the context of existing policy commitments. Second, they differ in the ubiquity and strength of parties claiming issue ownership on climate policy, with implications for party competition. In Denmark, the Socialist People's Party, the Social Liberals, the Red-Green Alliance, the Conservative People's Party, and since 2013 the Alternative, have all laid claim to being 'green' (Kosiara-Pedersen and Little 2016), while in Ireland political environment-alism is represented by a small Green Party. Third, climate change has been the subject of relatively little public concern in Ireland, while public con-cern has been greater in Denmark. Survey data show consistent gaps between the levels of concern in the two countries (Figure 1).

As similar cases, Ireland and Denmark allow us to focus on the roles of this set of factors in driving party preferences and ultimately generating cross-national differences in system-level climate politics. Although the countries differ in the configuration of climate policy preferences in their party systems and are diverse in their values for key causal factors, they are both *typical* cases of the expected relationship between left-right polariza-tion and climate policy polarization and as such they lend themselves to the examination of mechanisms that underpin the relationship between these factors (Seawright and Gerring 2008).

Case study methods and in-depth interviews are appropriate for exam-ining the mechanisms of interest. They potentially provide access to key actors' motivations and perceptions, which are central to understanding their responses to the incentives and constraints of interest. They allow us to follow preferences amidst a shifting policy agenda, and to examine how the causal factors of interest – which are potentially complementary and may interact – act in conjunction with one another.

We draw on a range of case study materials and the accounts of 22 individuals whom we interviewed between 2013 and 2016 (Table 1). Interviewees were selected because of their knowledge of individual parties (e.g. as elected representatives or advisers to ministers) or because of their knowledge of multiple parties on the climate policy issue (e.g. as

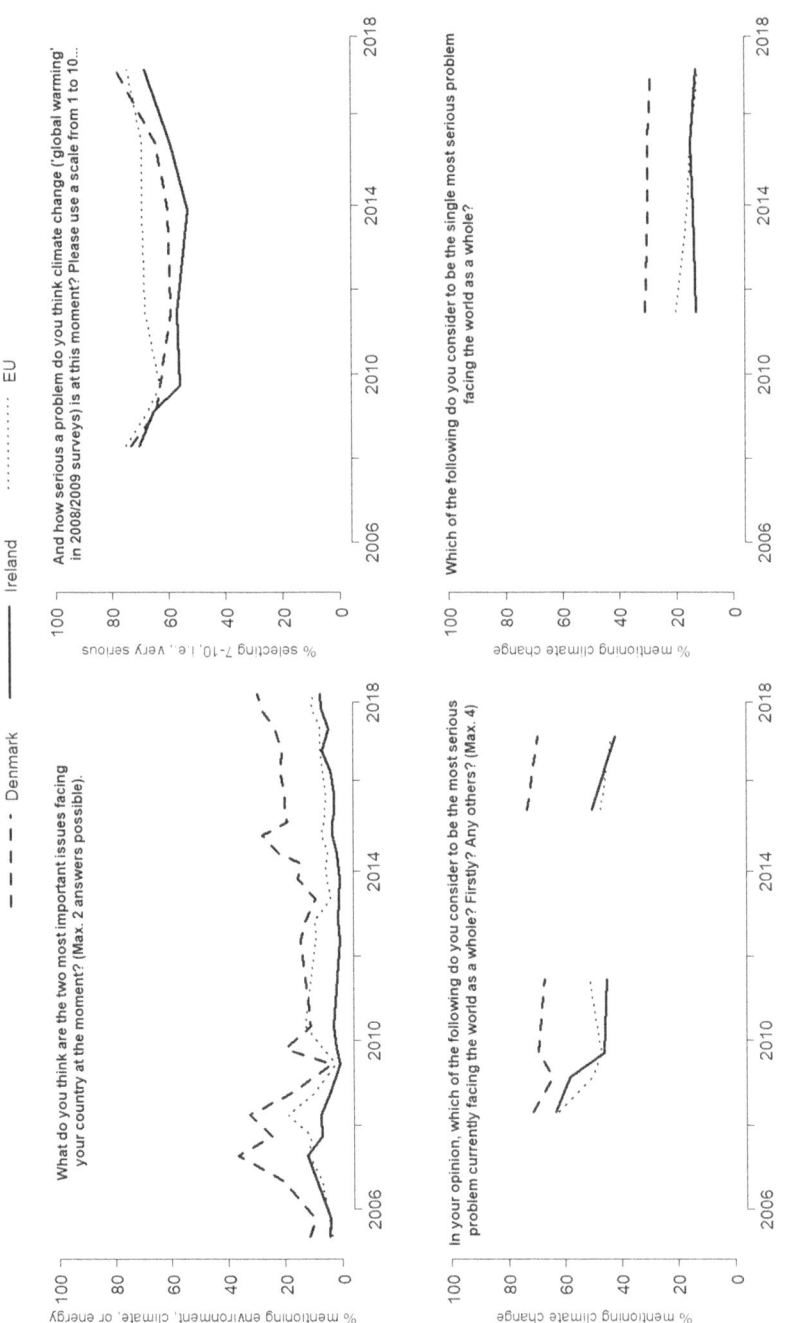

Figure 1. Public concern about climate change. (Source: European Commission 2018).

Table 1. Overview of interviews.

	Interview no.	Role	Month	Location
Ireland	1	Think tank representative	November 2014	Dublin
	2	NGO representative	December 2014	Dublin
	3	Political adviser	December 2014	Dublin
	4	Interest group representative	January 2015	Dublin
	5	Political adviser	July 2015	Dublin
	6	Political adviser	July 2015	Dublin
	7	Elected representative	June 2016	Dublin
	8	Political adviser	June 2016	Dublin
	9	NGO representative	June 2016	Dublin
	10	Elected representative	June 2016	Dublin
	11	Political adviser	June 2016	Dublin
	12	Interest group representative	June 2016	Dublin
Denmark	13	Elected representative	October 2013	Brussels
	14	Party leader	October 2013	Brussels
	15	NGO representative	January 2014	Copenhagen
	16	Elected representative	January 2014	Copenhagen
	17	Elected representative	January 2014	Copenhagen
	18	Elected representative	January 2014	Copenhagen
	19	Cabinet minister	January 2014	Copenhagen
	20	Interest group representative	January 2014	Copenhagen
	21	Party staff	March 2014	Phone interview
	22	Interest group representative	May 2014	Email

* Where the interviewee has held multiple relevant roles, we highlight the role that was most relevant for the purpose of this study.

representatives of NGOs or interest groups who lobbied the parties). The interviews focussed on the themes of office-, policy-, and vote-seeking incentives, and had a particular focus on gathering information about the internal party politics of climate policy development. They were conducted on a non-attribution basis and were not recorded. Relative to the overall period of interest, the interview data were collected during a short period, so the information from these sources is biased towards the ten years immediately prior to the interviews. This imbalance is partly offset by the publication, after the passage of time, of scholarly and journalistic accounts of the earlier part of the period of interest.

Case studies

Denmark, 1997-2016

Danish climate politics has typically been characterised by left-of-centre parties asserting issue ownership of climate policy and centre-right parties adopting a minimally accommodative policy stance (Interviews 16, 17, 21). This structure of partisan preferences has persisted, with the exception of the late-2000s when the centre-right aimed to outdo the left bloc on climate policy (Carter and Ladrech *et al.* 2018, pp.735–736). It consists less of a conflict over the need for climate policy than over the prominence of state involvement as well as the degree of international climate leadership Denmark should provide. These partisan differences have developed in spite of an underlying consensus on energy security, including leadership on renewable energy technologies and energy efficiency, which was forged in response to the Oil Crises of the 1970s and renewed in the development of subsequent policies, such as Energy 2000 (1990) (Andersen and Nielsen 2016, p.94, Eikeland and Inderberg 2016, p. 166).

Although climate change has not been a top policy area for public concern or party competition, it is considered part of the next tier of public attention, after health, jobs, and immigration (Kosiara-Pedersen 2008, Kosiara-Pedersen and Little 2016; Interview 16). Party attachment to climate policy is a party identity factor on the left and therefore parties on the left are expected to have more ambitious climate policy than parties on the right, even though traditional issues lead their campaigns. In addition to high public concern and the variety of potential issue-owners, environmental NGOs have been active (Binderkrantz 2015, p.126, Andersen and Nielsen 2016, p. 84), even providing the Social Liberal (SL) Minister for Climate and Energy from 2011 to 2015. Elements of Danish industry support ambitious Danish climate and energy policy, while traditional sectors such as agriculture have sided with the centre-right's less ambitious approach (Interviews 20, 22).

The Social Democrats' (SD) 'climate' identity was driven in part by activism of former party leader Svend Auken on the left of the party in the 1990s (Interviews 14–19) and was substantiated by government policies, including the adoption of an ambitious target for the first Kyoto commitment period. Under Prime Minister Poul Nyrup Rasmussen (SD) in 1993, the Environment and Energy ministries were merged, with Auken appointed its first minister; the integration of these two ministries 'created a synergy between the two policy fields which often have contradictory policy goals' (Dyrhauge 2017, p. 91). Together with SL, the Nyrup Rasmussen government consolidated Denmark's position as a climate leader. While in opposition from 2001 to 2011, the parties of the centre-left, including the SLs, continued to maintain pressure on the Venstre-led government (Seeberg 2016). Following significant gains at the 2007 general election (+12 of the *Folketing*'s 179 seats), the Socialist People's Party (SPP) signalled that it was ready to become a 'party of government', and together with SD developed a policy platform for the 2011 election in which climate and energy policy featured prominently (Interview 16). They aimed to outdo the government, including on its energy, transport, taxation and 'green growth' policies, leading not only to increased salience, but also refusal to support some government policies on the grounds of insufficient ambition (Laub 2012). While transitory, this competitive dynamic between the two main parties on climate policy was a powerful driver of party preferences (Seeberg 2016, Carter and Ladrech *et al.* 2018). The bargaining position of the SPP vis-à-vis SD, enhanced by its increased seat share, consolidated an ambitious climate policy position for the centre-left bloc.

Under Anders Fogh Rasmussen from 1998 and in government with the Conservative People's Party (CPP) from 2001 the centre-right Venstre prioritised industrial and agricultural interests and its climate policy was fitted into a neo-liberal framework; that is, the party promoted market-led developments rather than state-spending initiatives. In its first years in power, it highlighted conflict between climate policy and economic policy and its position entailed cutbacks on renewable energy projects, a retreat from leadership in the EU, and the appointment of the controversial Bjørn Lomborg to the new Environmental Assessment Institute (Andersen and Nielsen 2016, Eikeland and Inderberg 2016, Seeberg 2016). Fogh Rasmussen himself did not rank climate change as a priority issue, despite advice from his coalition partner to do so; he admitted as much at a 2008 party conference (Interview 14, 18).

Venstre changed orientation in the mid-2000s while in government, heralding a period of renewed convergence and higher-salience competition on climate policy between left and right. The high level of activity globally on climate policy influenced domestic politics and this period of salience was extended by the prospect of Denmark hosting the UN Climate Change Conference in 2009. Intra-bloc dynamics of party competition help explain this shift: the junior coalition partner (the CPP) maintained and slightly expanded its number of parliamentary seats at the 2005 election and kept them in 2007, whereas Venstre had lost seats in 2005 and 2007. These changes altered the weight of the CPP within the government coalition – where its party leader had continuously argued for a stronger climate policy (Interview 14), and thereby influenced government policy in a more active direction. Fogh Rasmussen appointed the popular Connie Hedegaard (CPP) as Environment Minister and established the environment as a 'top-five' priority; this was followed by the launch of a new energy policy in January 2007 with broad political support, the establishment of a Ministry for Climate Change and Energy, and an energy policy agreement in 2008 (Andersen and Nielsen 2016, p.86, Eikeland and Inderberg 2016, Seeberg 2016, p. 193). Developing Denmark's export potential was a major plank in Hedegaard's strategy and the about-turn in Venstre's position was justified by Fogh Rasmussen as enhancing energy security. In addition, significant sections of Danish industry and local (including municipal and private) interests had applied pressure to return to supporting renewable and decentralised sources of energy; these interests had in turn been created by earlier energy policies (Interview 15; Eikeland and Inderberg 2016).

From 2011 to 2015, the centre-left coalition maintained much of its climate policy ambition in government. Nonetheless, in straitened economic conditions, the SD finance minister presented the most significant intra-party obstacle to more ambitious policy, weighing short-term financial costs against longer-term investments (Interview 17) and the government reduced green taxes in an effort to increase economic growth (Andersen and Nielsen 2016, p. 92). Climate change was recognised as an issue 'owned' by the government parties, but the coalition also recognised that more traditional issues commanded the public's attention (Interview 16). Anticipating a departure from government after the 2015 election, the left focused its efforts on passing framework climate legislation (Interview 19).

Venstre's trajectory under Lars Løkke Rasmussen (from August 2009) was marked by retrenchment on climate policy and the prioritization of cost-cutting as a response to the economic crisis, leading to the re-establishment of clear differences between the parties. Both Fogh Rasmussen and Hedegaard had departed Danish politics in 2009 for NATO and the European Commission, respectively, thereby removing the two main climate leaders of the late 2000s from the centre-right. While Venstre entered into an agreement with the government on energy policy in 2012, it resisted elements of the government's climate and energy plan during its formulation, and of the right-of-centre parties only the CPP supported proposed framework climate legislation in 2014 (Kosiara-Pedersen and Little 2016, p.558; Interview 18). Venstre maintained that sufficient progress had been made on climate policy to focus elsewhere, a position also supported by the agricultural lobby but not by the peak industry association (Interviews 20–22). When Venstre returned to office in 2015, it pursued a decidedly less ambitious climate and energy policy (Burck *et al.* 2017, p. 19). The turnaround by Venstre is explained not only by the departure of Fogh Rasmussen and Hedegaard, but also by the electoral decline of the CPP, which suffered a significant loss of seats in the 2015 national election (-10), thereby easing the pressure on Venstre to maintain an ambitious climate mitigation policy. Venstre therefore depended all the more on parliamentary support from the Danish People's Party (DPP), which was unenthusiastic about climate policy and more attuned to agricultural interests.

Ireland, 1997-2016

Ireland's climate politics has been characterised by indistinct party preferences, especially between Fianna Fáil (FF) and Fine Gael (FG) (Interviews 2, 4, 7). The climate policy agenda has tended to focus on energy, on which there has been an absence of significant differences (Interview 8), and on agriculture, where a strong consensus has developed. This mainstream consensus was complemented by the absence of organised scepticism about climate science (Wagner and Payne 2017, p. 23). Low public concern with climate change has led to perceptions among party elites that climate policy has very low electoral value, with a partial and brief exception in 2006–2007 (Interviews 4, 5, 7, 8, 9, 12). Further, environmental NGOs have been weak (Interviews 6, 10). Arguments concerning competitiveness have regularly been used to counter climate policy proposals, such as a carbon tax and quantitative GHG emissions targets (e.g. Coghlan 2007, Interview 4, 11). Ireland's small size and irrelevance to the global problem of climate change is a theme that politicians across FF, FG, and Labour have aired at times (Interviews 5, 9).

There was some low-salience competition on environmental policy at the 1997 election, but without a particular focus on climate change. FF government minister Noel Dempsey initiated a period of policy entrepreneurship and the government adopted a modest but higher-than-expected target for the first Kyoto commitment period. This episode ended in 2004 with a failure to change FF's position on the key issue of a carbon tax in the face of blocking coalitions between intra-party actors, ministries, the junior coalition partner (the small, economically liberal Progressive Democrats), and interest groups (Coghlan 2007, pp.138–140, Cunningham 2008, pp.97–103, pp.146–151; Interviews 2, 5, 6, 8). Energy security concerns led the government to reject converting to gas Ireland's coal-burning power station at Moneypoint and to justify continued subsidisation of peat-burning for electricity generation (Forfás 2006, Cunningham 2008).

For a period of approximately nine months ahead of the May 2007 general election, the FF leadership sought to put climate change and the environment higher on its agenda, albeit this was not matched by positional changes on key issues such as carbon taxation or their plan to rely heavily on the purchase of carbon credits to meet Kyoto targets. There were at least three motivations for raising the salience of climate policy: competition with the Greens, who were expected to perform strongly in 2007; public opinion (see Figure 1), especially in key constituencies; and coalitional considerations, including increasing perceived compatibility with the Greens (Interview 5). Relatively high public concern was also recognised by candidates across several of the smaller, Dublin-centred parties who came to see it as a 'vote winner' (Interviews 2, 10, 11; Cunningham 2008, p. 106). When FF and the PDs entered government with the Greens in 2007, they effectively delegated aspects of climate policy to them (Interview 6). However, they also delayed the implementation of some important measures such as a carbon tax until it became necessary for revenue-raising reasons during the economic crisis. Their reluctance was at least partly due to the public's low willingness to pay the proposed tax

(Cunningham 2008, p.317, European Commission 2018: Special EB 300 and 322).

In opposition, FG did not make comparable efforts to emphasise climate policy. Internal policy entrepreneurship by Simon Coveney up to 2007 failed to convince the party of the merits of a carbon tax (Interviews 5, 6, 7). Although FG faced some of the same electoral incentives as FF, they had fewer reasons to improve their compatibility with the Greens, given the smaller party's clear preference for a FG-led coalition (Interviews 3, 5, 8). Labour also opposed a carbon tax due to concerns about costs for commuters and fuel poverty (Interview 11). With the Green Party's entry into government in 2007, Labour quickly sought to increase its focus on climate policy as a means of criticising the government while competing with its small but significant rival for middle-class votes, and driven by policy entrepreneurship of individuals such as its energy spokesperson Liz McManus (Little 2017a).

From mid-2008, economic crisis sent climate policy tumbling down the public and political agendas (Figure 1). The 2011 FF manifesto, published shortly after Ireland entered a multi-lateral 'bailout' programme, focused exclusively on jobs, growth and competitiveness, as did the FG agenda in government thereafter (Interview 5). Both main parties opposed the inclusion of binding GHG emissions targets in a new framework climate law (enacted in 2015) and important elements of climate policy were deprioritized during the crisis (Little 2017b). In government, FG not only opposed unilateral targets for Ireland, but also effectively disowned its EU emissions targets for 2020, blaming Ireland's failure to meet them on the unrealistic ambition of the previous government and reduced investment during the crisis. With no Green Party representation in parliament from 2011 to 2016, the junior coalition partner (Labour) helped to keep the issue of climate change on the political agenda (Interviews 9, 11).

The crisis increased the importance of economic sectors that could contribute to export-led growth and coincided with the strengthening and development of the cross-party consensus on agriculture, buttressed by industry-led national strategies for the sector. Although climate change was of particularly low salience among farmers in the 2000s, this changed in the 2010s (Interview 12), beginning with the activation of an IFA campaign against the Green Party's Climate Bill, which led to FF candidates fearing being 'outflanked' by FG on agriculture in rural constituencies (Interview 3). While in 2000, the FF-led government could envisage 'livestock reductions' as part of the first *National Climate Change Strategy* (NCCS 2000), by the 2011 general election the four main parties agreed that climate policy should not interfere with ambitious expansion plans for meat and dairy production, effectively putting this question outside the bounds of 'reasonable politics' (Interviews 9, 12). The decision of the

Labour Party, notably, to cleave to this consensus was a tactical one motivated by electoral and intra-party considerations (Interview 11).

Against this backdrop, the adoption of climate change legislation proposed by the FG/Labour government in 2015 was driven in part by Labour Party supporters' expectations of achievements on climate policy and by Labour's need to make good on its criticisms of the Green Party in government (Little 2017a, Torney 2017, Interviews 5, 11). The party's values, such as social justice and global responsibility, also help to explain why several Labour politicians engaged actively with the issue (Interview 11, 12).

Ireland's electoral system and political culture place a premium on local issues which, after 2010, provided a context for increased public concern about the local impact of various climate policies. Energy infrastructure (wind turbines and pylons) became a very significant issue across the main parties due to localised mobilization (Interviews 2, 6, 9, 11, 12), sometimes manifesting in intra-coalition and intra-party disputes (e.g. within the Labour Party on wind turbine setback distances and within both the coalition and FG on the North-South Interconnector) (Interviews 4, 9, 11). Another salient localised issue was the protection of peat bogs (which are important carbon sinks) at the expense of turf-cutting. Together with agriculture policy, these issues meant that parties' dependence on rural votes became increasingly associated with their positions on climate change. This dependence was greatest for the two main parties, but was also high for Sinn Féin and for Labour after its success in winning seats outside the main cities in 2011 (Interview 11).

Discussion

The causal factors of interest – public opinion, party competition, and existing policy commitments – frequently worked in conjunction with one another; nonetheless, in this section, we summarise and specify their roles. *Public opinion* was an important part of the context in which parties – especially mainstream parties – formed their policy preferences. In 2006–2008, the peak in public opinion in each country increased incentives for mainstream parties such as FF and Venstre to raise the salience of climate change and, in the case of Venstre, to develop 'greener' positions on climate policy, despite the second-order nature of climate policy as an issue. In the case of Ireland, low public concern was the reason interviewees most frequently cited to explain why politicians and their parties did not pay more attention to the issue.

The uneven distribution of public opinion across parties' potential support bases influenced their climate policy preferences and was in turn associated with left-right preferences. Party elites to the left-of-centre in both countries were aware that their supporters gave some priority to climate policy. Public opinion has also been mediated by institutions: in Ireland since approximately 2010, *localised public opinion*, in conjunction with constituency-specific competition among the main parties, has increasingly been a constraint on those parties' climate policy preferences in relation to various rural climate policy issues. While this is not a feature of Danish climate politics, there are comparable instances elsewhere (Stokes 2016). The urban-rural divide that these conflicts trace helps to explain some of the left-right differences that exist on climate policy.

We observe the power of mainstream *party competition* on climate policy in both directions. When the left bloc in Denmark increased their challenge to the right-of-centre government in the late 2000s, in the presence of policy entrepreneurship by the CPP and with the prospect of the Copenhagen Summit, there was powerful and short-lived competition between the main parties on climate policy. This did not take off in Ireland, partly due to a lack of credibility in FF's attempts to 'green' its image in 2007. What we do observe, however, is powerful competition between mainstream parties to oppose localised climate policies.

While we cannot resolve contradictory findings arising from broader cross-national studies on the role of issue owners, we have detected party preference changes motivated by strategies of accommodation found by Spoon *et al.* (2014). In Denmark, the SPP and CPP both influenced their respective blocs as they became larger in the 2000s; the Irish Greens most influenced other parties' preferences when they seemed likely to perform strongly in 2007. As expected, this competition has had uneven effects, with the greatest impact of competition from issue-owners on other left-of-centre parties. However, the Danish case also shows issue ownership and accommodation on the centre-right in the late 2000s when the CPP gained strength, as well as limited accommodation by FF of Green Party preferences in Ireland. Meanwhile, the DPP has acted as a competitive constraint on the Danish centre-right. Beyond electoral competition, concerns about coalitional considerations have also come into play when mainstream parties have perceived the need to develop their compatibility with issue-owners.

There is extensive evidence in our case studies that *existing policy preferences* influenced parties' climate policy preferences. These existing preferences were in turn rooted in members' expectations, which pushed parties of the left in particular to display a degree of ownership over the issue. In Denmark, state intervention was a major issue related to climate policy, while Venstre's neoliberal outlook in the late 1990s and early 2000s precluded certain climate policy positions. Likewise, the centre-right framed its turn to climate policy in the late 2000s in terms that resonated with its values, such as increasing exports and increasing energy security. For the centre-right parties in Ireland, competitiveness has been to the fore in arguments against climate policies, and energy security concerns have been used to justify both the continued use of fossil fuels and the development of renewable sources of energy. Economic policy preferences were prominent among the most influential existing policy preferences, suggesting that the 'economy vs climate' dimension remains relevant.

However, it is not simply the case that the left provided a context conducive to climate policy preference development while the right did not. In Denmark and Ireland, existing priorities of growth, financial stability, and social cohesion often trumped climate policy across the political spectrum, not least during the economic crisis. While the Irish Labour Party's support for climate policy was rooted in its broader values such as social justice and global responsibility, it also opposed the carbon tax due to concerns about fuel poverty and general consumer costs. The crisis also contributed to reshaping policy and party preferences on the left and right concerning Ireland's largest-emitting sector, agriculture, thus profoundly influencing climate politics. Indeed, evidence of the profound effects of the crisis on parties' preferences in both countries is striking (cf. Rohrschneider and Miles 2015).

The case studies also show that intra-party policy entrepreneurs sometimes have a key role in the formation of party preferences: Auken (SD), Hedegaard (CPP) and McManus (Labour). However, their success is not guaranteed, and these entrepreneurs had the greatest chance of success in influencing their own parties' policies in a context in which their initiatives coincided with other party goals (Little 2017a).

One other factor arising empirically from the case studies is that interest groups have evidently played a key role in shaping party preferences. The preferences of ostensibly similar interest groups vary, however, and the cases of Denmark and Ireland illustrate this quite clearly: the degree of support for climate policy from the main business interests differed across these contexts. However, agricultural interests remain a constraint on climate policy in both countries, consistent with the idea that sectors with high costs of mitigation will be most opposed to climate policy (Michaelowa 2000, Christoff and Eckersley 2011, p.442, Wagner and Ylä-Anttila 2018).

Overall, it is striking that in Denmark, where there is higher public concern, more competition from issue-owners, and a greater share of parties with left-of-centre preferences, this has not led to a uniformly 'greener' set of parties; rather, it has led to polarization at times, further indicating the positional, contentious nature of climate policies (Farstad 2018, p. 705). Curiously, the comparison of Denmark and Ireland does not fully support the widely-held view that partisan polarization of climate policy preferences is detrimental to climate policy: overall, Denmark has produced stronger climate leadership than Ireland over the period studied, despite being more polarized. This may be explained by an underlying consensus rooted in pre-existing policies: specifically, the aim of becoming a leader in energy technologies that developed after the Oil Crises, creating new constituencies of interest (Eikeland and Inderberg 2016). Nonetheless, when polarization has intensified in Denmark and the right has been in power, it has had an impact, as witnessed by its recent slide from a position of climate leadership (Burck and Marten *et al.* 2017).

Conclusion

Climate policy's characteristics make it an unusual and difficult issue for domestic political actors: it is cross-sectoral, it addresses a global problem, and it is associated with diffuse, distant benefits and concentrated costs. In the world of party politics, however, it also has much in common with other issues. When forming their climate policy preferences, parties respond to public opinion, they accommodate successful issue-owners, and they do not develop climate policies on a blank slate: they are constrained and facilitated by their existing policy preferences. The uneven effects of these factors across political parties help to explain differential responses observed across the left-right spectrum in many contexts. Given the centrality of political parties to policymaking, these findings have practical implications for climate policy practitioners.

While countries evidently vary in their climate politics, we might expect these mechanisms to be at work in other cases, at least in those typical of the relationship between left-right preferences and climate policy preferences. The strong role of pre-existing preferences may also be generalizable to party policies on other 'new' issues, as they are moulded to shape existing worldviews and policy commitments.

Our study has contributed to broadening and deepening the evidence-base on the drivers of parties' climate policy preferences. It has drawn on accounts of party elites' perceptions and motivations in responding to a set of factors that are important drivers of parties' preferences. It feeds into the growing body of knowledge on parties' climate policy preferences and comparative climate politics more generally, yet it also points to gaps in our knowledge.

The relationship between public opinion and party preferences still requires broad, crossnational studies that connect public opinion and parties, not least in Europe (Marcinkiewicz and Tosun 2015, p. 201). Our analysis focuses on party competition and climate policy preferences and in doing so it has highlighted the important role of mainstream party competition both as a catalyst for 'greener' climate policy preferences, but also as a constraint on party preferences. On intra-party factors, we show that existing preferences associated with traditional left-right politics matter for the policies that parties adopt and for the ways in which they frame them. Furthering these lines of enquiry with broader, systematic comparative studies will require building on recent studies (e.g. Carter and Ladrech *et al.* 2018, Farstad 2018) to develop more comprehensive and widely-applied measures of parties' climate policy preferences, especially of their *positions* on climate policy. Further, climate policy provides one cross-sectoral window through which relationships between party politics and interest groups can be viewed, expanding our knowledge of this key set of relationships (Binderkrantz 2014, p.535, 2015, p. 121).

Finally, our study highlights common drivers of climate policy preferences across two national contexts, but also suggests considerable diversity in the climate politics of small states. Characteristics that can most directly be related to state smallness – for example, perceptions that small size makes a country irrelevant to the problem of climate change – appear not to be universal, confounding fixed expectations about small states' climate politics.

Notes

1. Polarization is understood here as positional differences between parties. The opposite of partisan polarization is consensus (Dalton 2008, p. 909).
2. PI = SQRT$\{\Sigma(\text{party vote share}_i)^*([\text{party position}_i - \text{party system average position}]/5)^2\}$ where i is a party (Dalton 2008).
3. The data were collected for two Europe-wide Voter Advice Applications (Trechsel and Mair 2011, Garzia *et al.* 2017). Responses ranged from 'Strongly disagree' to 'Strongly agree' on the following statements: 'The promotion of public transport should be fostered through green taxes (e.g. road taxing)' and 'Renewable sources of energy (e.g. solar or wind energy) should be supported even if this means higher energy costs'. The Polarization Index values presented here are calculated using the aggregated values for these two items placed on a 0–10 scale.
4. Denmark's party system is also more fragmented than Ireland's, but fragmentation appears to be unrelated to polarization (Dalton 2008, p. 908).

Acknowledgements

This work was supported by the Economic and Social Research Council, Research Number ES/K00042X/1. We thank those who agreed to be interviewed in the course of our research. For their comments on earlier drafts, we thank two anonymous referees, Neil Carter, Detlef Jahn, Diarmuid Torney, Anders Wivel, and the participants in the workshop on Climate Politics in Small European States that was held at Dublin City University in June 2016.

Disclosure statement

No potential conflict of interest was reported by the authors.

ORCID

Robert Ladrech ⓘ http://orcid.org/0000-0001-6257-5051
Conor Little ⓘ http://orcid.org/0000-0001-5510-3195

References

Abou-Chadi, T., 2016. Niche party success and mainstream party policy shifts – how green and radical right parties differ in their impact. *British Journal of Political Science*, 46 (2), 417–436. doi:10.1017/S0007123414000155

Andersen, M.S. and Nielsen, H.Ø., 2016. Denmark: small state with big voice and bigger dilemmas. *In*: R. Wurzel, J. Connelly, and D. Liefferink, eds. *The European Union in international climate change politics: still taking a lead?* Abingdon: Routledge, 83–97.

Bakker, R., *et al.*, 2015. Measuring party positions in Europe The Chapel Hill expert survey trend file, 1999–2010. *Party Politics*, 21 (1), 143–152. doi:10.1177/1354068812462931

Båtstrand, S., 2014. Giving content to new politics From broad hypothesis to empirical analysis using Norwegian manifesto data on climate change. *Party Politics*, 20 (6), 930–939. doi:10.1177/1354068812462923

Båtstrand, S., 2015. More than markets: a comparative study of nine conservative parties on climate change. *Politics & Policy*, 43 (4), 538–561. doi:10.1111/polp.12122

Binderkrantz, A.S., 2014. Legislatures, lobbying, and interest groups. *In*: S. Martin, T. Saalfeld, and K. Strøm, eds. *The Oxford Handbook of Legislative Studies*. Oxford: OUP, 526–542.

Binderkrantz, A.S., 2015. Balancing gains and hazards: interest groups in electoral politics. *Interest Groups & Advocacy*, 4 (2), 120–140. doi:10.1057/iga.2014.20

Birchall, S.J., 2014. Termination theory and national climate change mitigation programs: the case of New Zealand. *Review of Policy Research*, 31 (1), 38–59. doi:10.1111/ropr.2014.31.issue-1

Brulle, R.J., Carmichael, J., and Jenkins, J.C., 2012. Shifting public opinion on climate change: an empirical assessment of factors influencing concern over climate change in the U.S., 2002–2010. *Climatic Change*, 114 (2), 169–188. doi:10.1007/s10584-012-0403-y

Burck, J., *et al.*, 2017. *The climate change performance index: results 2018*. Bonn/Berlin/Brussels: Germanwatch/CAN Europe.

Campbell, J.L. and Hall, J.A., 2015. Small states, nationalism and institutional capacities: an explanation of the difference in response of Ireland and Denmark to the financial crisis. *European Journal of Sociology*, 56 (1), 143–174. doi:10.1017/S0003975615000077

Carter, N., 2013. Greening the mainstream: party politics and the environment. *Environmental Politics*, 22 (1), 73–94. doi:10.1080/09644016.2013.755391

Carter, N. and Jacobs, M., 2014. Explaining radical policy change: the case of climate change and energy policy under the British Labour government 2006–10. *Public Administration*, 92 (1), 125–141. doi:10.1111/padm.2014.92. issue-1

Carter, N., *et al.* 2018. Political parties and climate policy: A new approach to measuring parties' climate policy preferences. *Party Politics*, 24 (6), 731–742. doi:10.1177/1354068817697630

Christoff, P. and Eckersley, R., 2011. Comparing State Responses. *In*: J.S. Dryzek, R. B. Norgaard, and D. Schlosberg, eds. *Oxford handbook of climate change and society*. Oxford: OUP, 431–448.

Coghlan, O., 2007. Irish climate-change policy from Kyoto to the carbon tax: a two-level game analysis of the interplay of knowledge and power. *Irish Studies in International Affairs*, 18 (1), 131–153. doi:10.3318/ISIA.2007.18.131

Compston, H. and Bailey, I., 2013. Climate policies and anti-climate policies. *Open Journal of Political Science*, 3 (4), 146–157. doi:10.4236/ojps.2013.34021

Cunningham, P., 2008. *Ireland's Burning*. Dublin: Poolbeg.

Dalton, R.J., 2008. The quantity and the quality of party systems. *Comparative Political Studies*, 41 (7), 899–920. doi:10.1177/0010414008315860

Dalton, R.J., 2009. Economics, environmentalism and party alignments: A note on partisan change in advanced industrial democracies. *European Journal of Political Research*, 48 (2), 161–175. doi:10.1111/ejpr.2009.48.issue-2

De Blasio, E. and Sorice, M., 2013. The framing of climate change in Italian politics and its impact on public opinion. *International Journal of Media & Cultural Politics*, 9 (1), 59–69. doi:10.1386/macp.9.1.59_1

Dyrhauge, H., 2017. Denmark: a wind-powered forerunner. *In*: I. Solorio and H. Jörgens, eds. *A guide to EU renewable energy policy*. Cheltenham: Edward Elgar, 85–102.

Eckstein, D., Künzel, V., and Schäfer, L., 2017. *Global climate risk index 2018*. Bonn: Germanwatch.

Eikeland, P.O. and Inderberg, T.H.J., 2016. Energy system transformation and long-term interest constellations in Denmark: can agency beat structure? *Energy Research & Social Science*, 11, 164–173. doi:10.1016/j.erss.2015.09.008

European Commission, 2018. *Various Eurobarometer surveys*. Available from: http://ec.europa.eu/commfrontoffice/publicopinion/index.cfm [Accessed 10 July 2018].

Farstad, F.M., 2016. *From consensus to polarization: what explains variation in party agreement on climate change?* York: University of York.

Farstad, F.M., 2018. What explains variation in parties' climate change salience? *Party Politics*, 24 (6), 698–707. doi:10.1177/1354068817693473

Forfás, 2006. *A baseline assessment of Ireland's oil dependence*. Dublin: Forfás.

Garzia, D., Trechsel, A., and L. de Sio, 2017. Party placement in supranational elections: an introduction to the euandi 2014 dataset. *Party Politics*, 23 (4), 333–341.

Gemenis, K., Katsanidou, A., and Vasilopoulou, S., 2012. The politics of anti-environmentalism: positional issue framing by the European radical right. Paper prepared for the *MPSA Annual Conference*, 12-15 April 2012, *Chicago*. doi:10.1094/PDIS-11-11-0999-PDN

Giavazzi, F. and Pagano, M., 1990. Can Severe Fiscal Contractions be Expansionary? Tales of Two Small European Countries. *NBER Working Paper No. 3372*. doi:10.1099/00221287-136-2-327

Guber, D.L., 2013. A cooling climate for change? Party polarization and the politics of global warming. *American Behavioral Scientist*, 57 (1), 93–115. doi:10.1177/0002764212463361

Hornsey, M.J., *et al.*, 2016. Meta-analyses of the determinants and outcomes of belief in climate change. *Nature Climate Change*, 6, 622–626. doi:10.1038/nclimate2943

IPCC, 2014. Summary for Policymakers. *In:* O. Edenhofer, et al., eds. *Climate Change 2014: Mitigation of Climate Change. Contribution of Working Group III to the Fifth Assessment Report of the Intergovernmental Panel on Climate Change.* Cambridge, United Kingdom: Cambridge University Press.

Jensen, C.B. and Spoon, -J.-J., 2011. Testing the 'Party Matters' thesis: explaining progress towards Kyoto Protocol targets. *Political Studies*, 59 (1), 99–115. doi:10.1111/j.1467-9248.2010.00852.x

Kluth, M. and Lynggaard, K., 2013. Explaining policy responses to Danish and Irish banking failures during the financial crisis. *West European Politics*, 36 (4), 771–788. doi:10.1080/01402382.2013.783358

Klüver, H. and Sagarzazu, I., 2016. Setting the Agenda or responding to voters? Political parties, voters and issue attention. *West European Politics*, 39 (2), 380–398. doi:10.1080/01402382.2015.1101295

Kosiara-Pedersen, K., 2008. The 2007 Danish general election: generating a Fragile Majority. *West European Politics*, 31 (5), 1040–1048. doi:10.1080/01402380802234730

Kosiara-Pedersen, K. and Little, C., 2016. Environmental politics in the 2015 Danish general election. *Environmental Politics*, 25 (3), 558–563. doi:10.1080/09644016.2015.1123825

Lachapelle, E. and Paterson, M., 2013. Drivers of national climate policy. *Climate Policy*, 13 (5), 547–571. doi:10.1080/14693062.2013.811333

Laub, L., 2012. *The struggle for the climate agenda: a discourse analysis of the Danish climate policy negotiations.* Copenhagen: CBS.

Little, C., 2017a. Intra-party policy entrepreneurship and party goals: the case of political parties' climate policy preferences in Ireland. *Irish Political Studies*, 32 (2), 199–223. doi:10.1080/07907184.2017.1297800

Little, C., 2017b. Portrait of a laggard? Environmental politics and the Irish general election of February 2016. *Environmental Politics*, 26 (1), 183–188. doi:10.1080/09644016.2016.1248617

Little, C. and Torney, D., 2017. The politics of climate change in Ireland: symposium introduction. *Irish Political Studies*, 32 (2), 191–198. doi:10.1080/07907184.2017.1299135

Marcinkiewicz, K. and Tosun, J., 2015. Contesting climate change: mapping the political debate in Poland. *East European Politics*, 31 (2), 187–207. doi:10.1080/21599165.2015.1022648

Meyer, T., 2013. *Constraints on party policy change.* Colchester, Essex: ECPR Press.

Michaelowa, A., 2000. The relative strength of economic interests in shaping EU climate policy: A hypothesis. *Energy & Environment*, 11 (3), 277–292. doi:10.1260/0958305001500130

NCCS, 2000. *National climate change strategy.* Dublin: Government of Ireland.

Rohrschneider, R., 1993. New party versus old left realignments. *The Journal of Politics*, 55 (3), 682–701. doi:10.2307/2131994

Rohrschneider, R. and Miles, M.R., 2015. Representation through parties? Environmental attitudes and party stances in Europe in 2013. *Environmental Politics*, 24 (4), 617–640. doi:10.1080/09644016.2015.1023579

Seawright, J. and Gerring, J., 2008. Case selection techniques in case study research. *Political Research Quarterly*, 61 (2), 294–308. doi:10.1177/1065912907313077

Seeberg, H.B., 2016. Opposition policy influence through Agenda-setting: the environment in Denmark, 1993–2009. *Scandinavian Political Studies*, 39 (2), 185–206. doi:10.1111/scps.2016.39.issue-2

Sohlberg, J., 2016. The effect of Elite Polarization: a comparative perspective on how party elites influence attitudes and behavior on climate change in the European Union. *Sustainability*, 9 (1), 39. doi:10.3390/su9010039

Spoon, -J.-J., Hobolt, S.B., and de Vries, C.E., 2014. Going green: explaining issue competition on the environment. *European Journal of Political Research*, 53 (2), 363–380. doi:10.1111/1475-6765.12032

Stokes, L.C., 2016. Electoral Backlash against climate policy: a natural experiment on retrospective voting and local resistance to public policy. *American Journal of Political Science*, 60 (4), 958–974. doi:10.1111/ajps.12220

Torney, D., 2017. If at first you don't succeed: the development of climate change legislation in Ireland. *Irish Political Studies*, 32 (2), 247–267. doi:10.1080/07907184.2017.1299134

Tosun, J., 2011. Political parties and marine pollution policy: exploring the case of Germany. *Marine Policy*, 35 (4), 536–541. doi:10.1016/j.marpol.2011.01.015

Trechsel, A.H. and Mair, P., 2011. When parties (also) position themselves: an introduction to the EU profiler. *Journal of Information Technology & Politics*, 8 (1), 1–20. doi:10.1080/19331681.2011.533533

Volkens, A., *et al.*, 2017. *The manifesto data collection. Version 2017b*. Berlin: Wissenschaftszentrum Berlin fur Sozialforschung (WZB).

Wagner, M. and Meyer, T.M., 2014. Which issues do parties emphasise? Salience strategies and party organisation in multiparty systems. *West European Politics*, 37 (5), 1019–1045. doi:10.1080/01402382.2014.911483

Wagner, P. and Payne, D., 2017. Trends, frames and discourse networks: analysing the coverage of climate change in Irish newspapers. *Irish Journal of Sociology*, 25 (1), 5–28. doi:10.7227/IJS.0011

Wagner, P. and Ylä-Anttila, T., 2018. Who got their way? Advocacy coalitions and the Irish climate change law. *Environmental Politics*, 27 (5), 872–891. doi:10.1080/09644016.2018.1458406

Creative and disruptive elements in Norway´s climate policy mix: the small-state perspective

Stefan Ćetković and Jon Birger Skjærseth

ABSTRACT

Recent scholarship has argued that effective and credible national climate policy mixes should encompass measures that promote new low-carbon technologies alongside those instruments aimed at constraining and phasing out support for existing polluting industries. The creative and disruptive policy measures in Norway´s climate policy mix are analysed by focusing on both national and international climate mitigation efforts. Norway´s climate policy mix at home has been more ambitious in the transport sector with a growing electric vehicle market than in the energy sector where niche support and disruptive policies have remained weak. Abroad, Norway has been increasingly active in supporting new low-carbon technologies and disrupting the fossil-fuel industry, especially coal. This is explained by the consensus-seeking and oil and gas dominated small-state social-investment political economy in Norway, combined with a forward-looking foreign policy based on norm-setting and multilateralism.

Introduction

As global attention towards mitigating climate change intensified following the Paris Agreement in 2015, so has the scholarly interest in national climate policies, their design and implementation challenges (Peters *et al.* 2017, Schoenefeld, Hilden and Jordan 2018). There is increasing understanding that the substantial transformation towards a low-carbon economy cannot be achieved with single policy interventions, but rather through a well-designed mix of mutually reinforcing policy measures (Rogge and Reichardt 2016). Recent scholarship has convincingly argued that effective and credible national climate policy mixes should encompass measures that promote new low-carbon technologies and sectors alongside instruments aimed at constraining and phasing out support for existing polluting industries (Kivimaa and Kern 2016, Geels *et al.* 2017). The conceptualization of climate policy mixes targeting ´niche support´ and ´creative destruction´ and how the scope and effects of such policy measures are shaped by broader political-economic factors is, however, yet to be fully developed and understood.

Here, we seek to advance this debate by investigating the ´niche support´ and ´creative destruction´ elements in the climate policy mix in Norway over time. Furthermore, we explore how political-economic and foreign policy features of Norway as a small, advanced and corporatist economy have influenced the relative weight of different elements in the climate policy mix. Empirically, we add to the existing literature by providing an empirically-rich account of the key ´niche support´ and ´creative destruction´ climate policy measures in Norway and their historical development. As a theoretical contribution, we further develop the conceptual framework for scrutinizing ´niche support´ and ´creative destruction´ climate policy instruments by proposing a more consistent and comprehensive framework for analysis. We also demonstrate how the small state-specific features of Norway´s political economy and foreign policy provide valuable insights into the factors that shape the creative and disruptive character of the climate policy mix. With this, we aim to address the gap in the literature as few studies have drawn on the insights from comparative political economy to explain national climate politics and policy (but see Lachapelle and Paterson 2013). By incorporating the foreign policy factor, we seek to contribute to the literature on the role of international affairs and symbolic politics in national climate policymaking (Harris 2002).

With its long history and remarkable activity in climate policy at home and abroad, Norway represents a pertinent case for studying climate policy design and policy change. Norway is a small European country, if measured by its direct contribution to global greenhouse gas (GHG) emissions, which lies around 0.1% of the global total. However, this figure hides the actual impact, capacity and responsibility of Norway with respect to climate change. Norway is the second-wealthiest OECD country in GDP per capita (OECD 2016), and most of this wealth comes from the production and export of carbon-intensive fuels: oil and gas. Norway is the world's fifteenth largest oil producer and the sixth largest producer of gas. Around 16% of Norway's GDP and 40% of its exports stem from the petroleum sector excluding the service and supply industry (Norwegian Petroleum 2019). Moreover, Norway is ranked the seventh largest exporter of GHG emissions globally (Oil Change International 2017). Not only does Norway hold considerable historical responsibility for driving global climate change, but it has also continuously expressed its commitment to contribute to mitigating climate change. The recent focus on combating climate change and decarbonizing the economy, at the EU and international level (via the

'Paris Agreement'), has placed even greater demands on Norway to re-think its climate policy and diversify its dependence on fossil-fuel-based revenues.

Against this background, we explore two sets of questions. First, what policy mix has the Norwegian government employed to reduce GHG emissions, and how has that mix changed over time? The policy mix is understood as a combination of nationally defined climate-policy goals and means encompassing ´low-carbon technology creation´ and ´fossil-fuel destruction´ strategies. Second, what factors help explain the relative weight of creative and destructive elements in Norway's climate policy mix, and their stability and change over time? In addressing the first question, we conceptualize creative and destructive climate policies and their historical evolution by building upon the framework of 'niche support vs. creative destruction' outlined by Kivimaa and Kern (2016). We focus on climate mitigation measures in the energy and transport sectors (excluding aviation) as the largest GHG emitters in Norway (Statistics Norway 2017). In answering the second question, we draw on insights from studies of the comparative political economy and foreign behaviour of small states as important variables in explaining national policy responses to climate change.

Towards a theoretical framework

In the subsequent sections, we develop our theoretical framework for conceptualizing and explaining the creative and disruptive features of the climate policy mix in Norway. In the first section we outline the framework for capturing the nature and evolution of the climate policy mix by drawing on the insights from the studies on sustainability transitions and climate policy innovations. In the second section, we briefly review the literature on the political-economy and foreign policy of small states, and Norway in particular. This will allow us to identify the key concepts and theoretical assumptions concerning the relationship between Norway´s small-state characteristics and climate policymaking.

Conceptualizing climate policy mix: between creation and destruction

Despite growing investments and significant cost-reductions in renewable energy and other low-carbon technologies, global GHG emissions have failed to decrease substantially (IEA 2017). This raises the question of whether the stagnation or even rise in GHG emissions (Globalcarbonproject 2017) is a temporary feature of generally sensible low-carbon transitions or whether it is a sign of ill-defined policies, lack of political resoluteness and carbon lock-in. Scholars have recently suggested the stronger focus on the disruptive character of sustainability policies.

They have argued that a genuine transition can only be achieved if policy strategies for promoting low-carbon technologies are accompanied by credible policy efforts to constrain and eventually phase-out polluting industries and practices. Uncovering the creative and destructive elements of climate policy mixes thus offers important insights into the comprehensiveness and credibility of national decarbonisation strategies. The investigation of creative and destructive sustainability policy measures is still at the early stage. Kivimaa and Kern (2016) have proposed a framework for categorizing creative and destructive policy measures and applied it to understand the energy efficiency policy mix in Finland and the UK. They conclude that the policy mix in both countries is unbalanced in favouring niche creation over creative destruction. David (2017) has applied a similar approach to the energy transition in Germany, labelled ´innovation vs. exnovation, innovation referring to the creative part of the policy mix whereas exnovation refers to measures to destabilize the fossil-fuel regime. He established that the policy mix is quite developed but suffers from inconsistency and incoherence, particularly on the exnovation side, which has hampered the decarbonisation efforts. By focusing on the interplay between destructive and creative policy instruments in the context of the German energy transition, Rogge and Johnstone (2017) found evidence of the positive effects of the destructive policy instruments and the phase-out of nuclear energy on technological change and innovations in the emerging renewable energy sectors. Despite the valuable theoretical and empirical insights contributed by the existing literature, there is a lack of consensus on how to classify and conceptualize the key policy instruments for niche creation and creative destruction to allow for more consistent investigation and comparison across national contexts.

Our investigation of the climate policy mix in Norway is inspired by the framework of Kivimaa and Kern (2016) but we modify and hopefully improve the framework in several important respects. First, rather than focusing on the quantity of all adopted policy instruments at a given time, we discuss the most important policy measures and their historical evolution. In so doing we seek to capture the stability and change of the policy mix over time, but also provide a more detailed account of the character and impact of different policy measures. Second, we include in the analysis not only domestic climate policy efforts, but also Norwegian policy measures aimed at promoting new technologies and disrupting established fossil-fuel industries internationally. This serves to enrich the ´creation vs. destruction´ analysis and emphasize the scope and flexibility of policy instruments available to national governments to mitigate climate change. Third, we streamline the framework of Kivimaa and Kern by suggesting a focus on four main policy categories in both niche creation and creative destruction dimensions (see Table 1).

Table 1. Taxonomy of niche support and creative destruction measures.

	Niche support	Creative destruction
Energy production /Transport	**Market formation** *Incentives for promoting demand for new low-carbon technologies*	**Control policies** *Market and regulation-based restrictions on polluting sectors*
	Support for low-carbon R&D *Policy measures aimed at supporting the development and demonstration of new low-carbon technologies*	**Constraints on carbon-based R&D** *Decrease in public spending on RD&D for carbon-based sectors*
	Resource mobilization for low-carbon sectors *Measures for mobilizing capital for low-carbon projects and technologies*	**Constraints on carbon-based investments** *Decrease in government subsidies and investments in carbon-based business operations*
	Strategic goals for developing low-carbon technologies *Time-frames adopted for expanding new low-carbon sectors and technologies*	**Strategic goals for constraining carbon-based technologies** *Time-frames adopted for constraining and phasing out carbon-based sectors and technologies*

Source: Adapted from Kivimaa and Kern (2016)

Political economy and foreign policy of small states: the case of Norway

What can the institutions of the small advanced economy of Norway tell us about the likely nature and evolution of its climate policy mix? We argue that two dimensions of small states are particularly important for understanding their climate policy strategies: corporatist political economy and ambitious forward-looking foreign policy.

Small-state political economy of climate policy in Norway

Although not without its shortcomings, Katzenstein's (1985) seminal work on small states in world markets remains the chief reference point for understanding the politics and policy of small states (Ingebritsen 2010). In more political-economic terms, Keating and Harvey (2014) distinguish between the market-liberal and social-investment models among small states. Keating (2015) argues that the success of small, open economies hinges on two critical conditions. The first is access to external markets – essential for compensating for a small domestic market and low economic diversification. Such economies are often keen proponents of trade liberalization and non-discriminatory economic measures globally. Although not a full EU member, Norway has participated in the internal EU market since 1994 through the Agreement on the European Economic Area (EEA), which grants Norway access to the single market, conditional on Norwegian implementation of EEA-relevant EU legislation. Norway has become increasingly integrated with the EU mainly because the EU is seen as the major market for Norwegian products, especially natural gas. Around 75% of Norway's trade is with the EU while almost all Norwegian gas exports are directed to the EU market (European Commission 2017). As a non-EU member, Norway has limited formal influence over EEA-relevant EU legislation, including the ever expanding EU climate and energy polices. Norway has, however, used the available institutional and informal channels to inform EU energy policy and mediate its domestic effects (Hofmann et al. 2019).

The second condition for success is the existence of a consensus-based institutional setting combined with extensive public spending, serving to buffer and enable adaptation to volatile economic forces. Katzenstein (1985) describes such institutional arrangements as 'democratic corporatism'. Consensus-seeking and partnership-based policymaking arrangements tend to facilitate long-term policy planning and coordination, but may impede deeper reforms and neglect larger problems (Keating and Harvey 2014). It is worth noting that corporatism is not an exclusive feature of small states, as larger economies can also entail corporatist state-market structures (Hall and Soskice 2001). The democratic corporatism in small advanced economies is, however, distinctive due to closer state-market ties and a more inclusive policy process, which is also a result of a smaller population and often a unitary state structure. Existing research shows that corporatist structures in Norway have facilitated stable and long-term environmental and energy policies, but have prevented more radical and disruptive policy solutions (Dryzek 2003, Ćetković, Buzogany and Schreurs 2017). While all small economies of the social investment model share some common features, they are not uniform in their adaptation strategies. These depend on many factors, including elite beliefs, social identity and the relative weight of different national economic sectors (Ingebritsen 2010). A closer look reveals at least four key elements that characterize the Norwegian political-economic landscape: first, a trade-oriented economy based on energy-intensive industries, exploitation of natural resources and close collaboration between established industrial clusters and domestic research institutes; second, a social welfare model based on comprehensive state–capital–labour wage bargaining, high taxation, low inequality and strong emphasis on balanced regional development; third, a proportional representation electoral system with a consensus-seeking policy style; fourth, an influential role for pragmatic thinking and the economics-dominated academic community (Fagerberg *et al.* 2009, Mjøset and Cappelen 2011, Dyrstad 2015). Since the first major discovery of oil in 1969, the petroleum sector has evolved into a cornerstone of the Norwegian welfare state. The electricity sector is almost entirely dependent on carbon-free hydropower, which reduces the domestic climate mitigation options to the transportation and oil and gas sectors.

From this literature, we expect a consensus-oriented and coordinated-market economy such as Norway's to show considerable capacity for the development and adjustment of long-term climate policy instruments, not least because of the need to implement EU energy and climate policy and adapt its export-oriented economy to more exacting GHG emission standards. Further, strong petroleum sector corporate influence and tight and consensus-seeking policymaking should favour incremental niche support rather than radical destruction. In addition, more disruptive policy measures could be expected in the transportation sector given the considerable mitigation potential, lack of strong domestic corporate interests and the diffuse character of vehicle emissions, which makes policy change politically easier to implement.

Small-state foreign policy of Norway and climate policy

The importance of the national foreign policy strategy for the country's climate policy is an important but largely neglected issue (for exceptions, see Harris 2002, Cass 2008). Although foreign policies of small states have traditionally not received as much attention as those of great powers (Neumann and Gstöhl 2004), scholars have begun showing interest in the subject (Ingebritsen *et al.* 2006, Björkdahl 2008). The literature views foreign policy of small states as a product of structure or agency, or a mixture of both (Neumann and Gstöhl 2004). The structuralist approach emphasizes the rules and norms rooted in the dominant material and geo-strategic relations in the international system which impact on national foreign policy strategies. Regarding foreign-policy behaviour of small advanced economies, several propositions from the structuralist approach can be formulated. Small states are more interested in maintaining and promoting international law as a means of ensuring national security and compensating for their low military power. Here, the governments of small states typically employ soft power, norm-advocacy and reputation as key foreign policy instruments (Björkdahl 2008). With their constrained human and material resources, they often focus on a few 'progressive' foreign policy areas, such as environmental protection and peaceful conflict resolution. This is further related to the need and propensity of small advanced economies to adopt a proactive foreign policy stance for better control over agenda-setting in the international arena (Neumann and Gstöhl 2004). Whereas structural factors constitute important explanatory variables of foreign relations, the foreign policy of small states can be properly understood only when specific national characteristics are considered. These include material interests, institutional settings as well as elite and societal ideas and beliefs (Gvalia *et al.* 2013).

In line with the theoretical expectations, Norway's foreign policy has been part of the 'Nordic exceptionalism' characterized by a strong presence in multilateral institutions, high development assistance spending and keen support for environmental and social concerns (Hansen and Gjefsen 2015). In the words of the Norwegian Minister of Foreign Affairs, 'Democracy, human rights, sustainable development and an international legal order form the basis of our foreign and development policy' (Brende 2015). Norway's active foreign policy and safeguarding of international rules and norms have been motivated largely by concerns for its own security and vulnerability (Norwegian Ministry of Foreign Affairs 2009, Pauly and Jentleson 2014). Beyond such interest-driven behaviour, concerns about international justice and peace seem deeply entrenched in Norway's identity and self-image (Skånland 2010). Sustainable development and environmental protection have constituted important elements of Norway's foreign policy ever since the former prime minister, Gro Harlem Brundtland, chaired the UN World Commission on Environment and Development. The ambition of being a global environmental leader spilled over to the issue of climate change (Eckersley 2016). Norway's active role in climate-change negotiations has been motivated not only by the concerns for the country's international reputation as a norm-setter, but also by the desire to influence climate change agreements in line with Norwegian preferences and interests. Being a small, open, petroleum-based economy, Norway has advocated flexible, market-based climate policy solutions together with technologies that enable further use of fossil-fuels. An example of the latter is Norway's support for carbon capture and storage (CCS) (Roettereng 2016).

Against this background, we would expect Norway to maintain its ambitiousness in climate policy goals. While this should positively affect domestic climate mitigation efforts, Norway should be less constrained in mitigating climate change abroad than in its efforts to reduce GHG emissions at home, given strong vested interests and high abatement costs. Overall, Norway´s climate policy mix is likely to particularly feature climate mitigation measures abroad, possibly targeting both niche creation and creative destruction.

Methodology

To identify the main climate and energy policy measures in Norway we used the IEA database and complemented it with analyses of official documents published on the websites of the ministry departments and relevant agencies. We also consulted existing scholarly analyses, expert reports and media coverage to triangulate the data. In the analysis of the effects and drivers of the identified policies, we follow the theory-guided process-tracing method, which relies on established theoretical propositions to describe and explain the dynamic and interrelated role of institutions, ideas and interests in the policy process over time (Falleti 2016). Specifically, we employ a historical institutionalist perspective to trace the relationship between the adopted policy measures and actors, structures and processes at the level of national political economy and climate diplomacy of Norway.

Niche support and creative destruction in Norway's climate policy

In this section we describe the major policy actions that Norway has taken to address climate mitigation, breaking them down into two main categories: niche support and creative destruction policies. In addition to describing the character, evolution and impact of different policy instruments, we also analyze the drivers behind the adopted measures by employing the political-economic and foreign policy insights on Norway as a small, advanced, social-investment economy.

Niche support

Market formation

Given the projected increase in energy demand and shortage of new power-production capacities, Enova was established in 2001 to support the realization of new renewable energy-based production facilities. Since the focus was on cost-effective energy supply, only advanced technologies were supported. Onshore wind, for instance, came into focus only during 2008–2010. Enova funding was soon replaced by a green certificate scheme as the main renewable energy support mechanism. In 2012, Norway joined a green certificate system with Sweden, setting the 2020 goal of reaching 28.4 TWh in new renewable energy in both countries combined. Although Norway and Sweden share the costs of the support scheme almost equally – Sweden is financing 15.2 TWh and Norway 13.2 TWh – by 2016 Sweden reached 14.34 TWh and Norway only 3.43 TWh in renewable electricity supported by the scheme (NVE & Energimyndigheten 2016). Green certificates are a technology-neutral instrument favouring mature technologies (e.g. hydropower). In 2017, the Norwegian and Swedish governments agreed to extend the green certificate scheme until 2030 but only for Sweden. Norway decided not to commit to new targets after 2020 (Norwegian Ministry of Petroleum and Energy, 2017).

Another market potential for renewable energy involves replacing the carbon-based power supply of offshore oil and gas installations with low-carbon renewable electricity (Blindheim 2015). Electrification can be sourced from land (onshore wind and hydropower) or through offshore wind power installations. Since 1996, companies have been mandated to consider electrification in connection with the licensing process, and the 2012 White Paper on Norwegian climate policy emphasized increasing the electrification of petroleum fields (Norwegian Ministry of the Environment 2012). However, with a new centre/right coalition in power, the 2015 White Paper dropped this requirement (Norwegian Ministry of Climate and Environment 2015). In 2013/2014, there was a parliamentary debate on requiring full-scale electrification from the start for the major new North Sea oil project 'Johan Sverdrup'. Eventually, mandatory electrification of the project was postponed until 2022. Despite the rise in state R&D spending, broader legitimacy and a functioning market for new low-carbon technologies such as offshore wind and CCS have failed to develop (Normann 2014, 2015)

The key instrument for reducing GHG emissions in the transport sector has been purchase incentives for electric cars. These incentives have been gradually introduced over the past two decades, but the concerted government action and rapid increase in electric vehicles started in 2009 (IFE 2015). Passenger cars are heavily taxed in Norway, whereas electric vehicles are either fully exempted or subject to reduced tax rates. Electric vehicles are also allowed to drive in bus/taxi lanes and enjoy free public recharging stations. These measures have been successful in creating one of the most dynamic markets for electric vehicles in the world: by April 2015, Norway had more than 50,000 electric cars (IFE 2015). Promoting electric vehicles is an attractive climate policy instrument due to Norway's dominantly hydro-power-based zero-carbon electricity production. Although the reductions in electric vehicle subsidies have been debated, the decision has been made to extend the support scheme until 2020 (Norwegian Ministry of Climate and Environment 2017).

Overall, creating the market for established renewable energy technologies through the green certificate scheme has been tied to Norway's climate foreign policy and political and economic linkages with the EU, but progress has been slow and Norway has accepted fulfilling its green electricity targets mainly by financing projects in Sweden. In the transportation sector, weaker incumbent actors and interests have enabled Norway's success in creating the lead market for electric vehicles and so reducing its GHG emissions and strengthening its international image as a climate policy leader.

Support for low-carbon R&D

State support for research on energy and transport technologies began with a technology-neutral approach but eventually included more targeted measures to promote infant low-carbon technologies. State R&D funding for clean energy and transport reached a turning-point in 2008 following a broadly based political compromise ('the climate agreement') reached among all major political parties in the parliament, except for the neoliberal Progress Party (Government of Norway 2008). In the same year, a national energy R&D strategy 'Energi21' was announced, with the overarching vision of strengthening Norway's status as 'a climate-friendly energy nation'. The main novelty of Energi21 lies in its focus on promoting both mature and infant climate-friendly energy technologies where Norway has expertise and potential comparative advantages (ENERGI21 2014). The previous energy research programme, RENERGI, expired in 2013 and was replaced with ENERGIX, more closely aligned with Energi21. Since 2005, a programme for research on CCS has also been in place.

Enova, Norway's main energy funding agency, has worked to promote the demonstration and testing of infant and close-to-market technologies, such as offshore wind (ENOVA 2015). Funding demonstration and testing of clean transport technologies became the responsibility of Transnova, a state enterprise established in 2009. Its budget varied from NOK50 million in 2009 and 2010 to NOK74.8 million in 2012. Transnova was instrumental in creating the infrastructure for electric vehicles through financing free public recharging stations.

Although the government share of R&D in low-carbon technologies has increased sharply since 2008, stabilizing in recent years, overall R&D spending has actually declined. Total state and private R&D spending on renewable energy recorded a sharp decline in 2009–2013, due to lower private-sector involvement, as did that on CCS, where spending declined after peaking in 2011 (Research Council of Norway 2015). The increase in the R&D support was clearly associated with EU renewable energy policy and the binding national targets formulated in 2009. Norway's participation in the EEA has stimulated policy change towards more R&D support for niche technologies, but the unfavourable political-economic conditions reflected in the resistance of incumbent actors and low electricity prices have prevented the creation of the market for such technologies.

Resource mobilization for low-carbon sectors

The support for climate-friendly business has been channelled mainly through the state-owned enterprise 'Innovation Norway'. The budget of its Environmental Technology Programme has increased considerably since its creation in 2010, from NOK140 to 465 million in 2016 (Innovation Norway, n.d.). Exports of Norwegian companies, including those operating in clean energy and transport, are supported through 'Export Credit Norway', but the overall share of support secured by these industries is marginal. In 2014, the wind power industry received the most support, but amounted to only 1.13% of the overall budget (Export Credit Norway 2014). Strong state ownership in two major energy utilities, Statkraft (100%) and Statoil (67%), gives the government some influence in directing investments to priority energy sectors and projects. Although the two companies operate according to market rules, there has been increasing pressure on energy utilities to support the government's efforts in climate-friendly technologies. In 2014 the parliament voted to support Statkraft with NOK 5 billion; an additional NOK 5 billion should come from reducing the dividend to the state in the period 2016–2018 (Statkraft 2014). The government expected Statkraft to invest in renewable energy projects. However, prior to the adoption of the 2016 budget, the government cut its support by reducing dividends, which forced Statkraft to abandon planned offshore investments and to reconsider the business model for several hydropower and onshore wind projects (Statkraft 2015).

Alongside domestic resource mobilization, Norway has played an important role in supporting climate change policies and technologies abroad. As one of the largest development aid donors, Norway has declared environment and energy as key priority areas in its development assistance. The previously stable development assistance budget for environment and energy hit a record high in 2008 and again in 2013, following the two national parliamentary climate agreements from 2008 and 2012 (Government of Norway 2014b). As an EEA member, Norway supports the development of less developed regions and countries in the EU. Around one third of the entire budget is dedicated to environmental protection and climate change (EEA Grants n.d.). Norway is also investing to reduce deforestation in developing countries. At the 2007 Bali climate conference, Norway launched a major deforestation programme, pledging to contribute NOK3 billion annually until 2015 to counter deforestation. In 2011, reduced deforestation in Brazil apparently amounted to 10–20 times Norway's annual GHG emissions (Norwegian Ministry of the Environment 2012), but some have queried whether this was due to the funds allocated. Ahead of the Climate Conference in Paris in 2015, the government made the decision to scale up its contribution to the Green Climate Fund to increase the prospects for reaching a global climate agreement (Government of Norway 2015).

The considerable investments in mobilizing resources for low-carbon technologies and climate change mitigation at the international level is in line with Norway's foreign policy concerns for its international reputation and interests in stable global climate agreement. Domestically, it has been more challenging to financially support low-carbon investments given the lack of bottom-up pressures but also due to low oil prices which constrained public finances.

Strategic goals for developing low-carbon technologies

The first comprehensive objective for increasing Norway's share of renewable energy sources was adopted in 2012, transposing the requirement from the EU Renewables Directive. The national target is to increase the share of renewable energy consumption from 60.1% in 2005 to 67.5% by 2020 (IFE 2015). Further, the government aims to implement at least one full-scale CCS facility by 2020 (Norwegian Ministry of Climate and Environment 2015). Norway has no current formal targets for expanding its electric vehicle market, but it has recently set the target of zero-emission for all new passenger vehicles and light vans from 2025 (see section on strategic goals for creative destruction below).

Norway has reluctantly transposed EU renewable energy policy through a binding national target. Domestically more important in political and economic terms is the development of CCS, which if commercialized would enable fossil-fuel industry to continue operating with a lower climate change impact. Building on the success in mainstreaming electric vehicles, the government has further raised its targets for decarbonizing transport as an important instrument for meeting national climate commitments.

Creative destruction control policies

Carbon pricing based on the cost-effectiveness principle has been Norway's overriding climate policy instrument since the early 1990s. Alongside the general purpose of revenue raising, the CO_2 tax has been designed to stimulate less carbon-intensive oil and gas extraction and to promote the use of low or zero-carbon transport models. The CO2 tax covers about 55% of domestic emissions, with tax levels varying from about €3 to almost €50 per tonne. Petrol is subject to the highest tax rate, and land-based consumption of gas the lowest rate. The largest documented effect of the tax has been in the petroleum sector. It has, for example, facilitated CO2 storage at Norway's Sleipner gas field, amounting annually to about 1 million tonnes of CO2 since 1996 (Skjærseth and Christiansen 2006). In 2013, the carbon tax on offshore oil and gas was doubled, from NOK210 to NOK410 per tonne. The CO2 tax was supplemented by a domestic emissions trading system in 2005, encompassing some 10% of emissions not covered by the CO2 tax. From 2008, the Norwegian system became fully integrated in the EU Emissions Trading Scheme. Although some studies have presented an economic and environmental rationale for addressing the supply-side of GHG emissions by reducing oil and gas extraction (Fæhn *et al.* 2013), such proposals have been dismissed by the government. In addition to the CO2 tax, Norway employs stringent emission standards for new motor vehicles in order to curb GHG emissions from transport.

Given the long-term oriented and consensus-driven policymaking in combination with active foreign climate policy, Norway has continued to rely on the carbon-tax as a key control policy. The CO2 tax is not designed to structurally disrupt fossil-fuel industries, but to increase the environmental efficiency of the oil and gas companies and stimulate incremental innovations. This approach can be explained by the small-state character of Norway related to its corporatist structures and the government role in helping domestic companies to adapt to changing political and market conditions.

Constraints on carbon-based R&D

There is little evidence on the withdrawal of support for R&D in petroleum research; indeed, state R&D funding was strengthened with the introduction of a large-scale PETROMAKS research programme in 2004. The annual budget for petroleum research varied in subsequent years, but recently overall public spending on petroleum R&D has been further institutionalized with a steady increase in the allocated funds. In 2013, the large-scale Petromaks 2 programme was launched, with a larger budget compared to the first Petromaks (ERKC n.d.). The central objective of petroleum R&D is to facilitate effective and environmentally sound extraction from remaining oil fields.

Overall, Norway has maintained support for R&D in the oil and gas industry in line with the established interests and the government policy of protecting and economically relying on the innovative oil and gas companies.

Constraints on carbon-based investments

The government has enhanced its efforts to promote the phase-out of subsidies for fossil fuels internationally. In 2010 Norway joined an initiative of several non-G20 countries, 'Friends of Fossil Fuel Subsidy Reform'. In 2014, the government adopted a national strategy for reforming international fossil fuel subsidies (Government of Norway 2014a). Another important initiative has been the 2015 decision of the Norwegian Government Pension Fund, the largest sovereign wealth fund in the world, to divest itself of shares in companies that have more than 30% of their portfolio in the coal business (Carrington 2015).

These efforts to reduce fossil-fuel subsidies globally stand in stark contrast to the extensive government subsidies and ambitious extraction policy for domestic oil and gas. Government subsidies for upstream oil and gas activities in Norway in 2009 were estimated at around US$4 billion (GSI & IISD 2012). The government has also continued to support the expansion of oil and gas drilling in new areas in the Arctic zone. In 2016, it awarded ten new licenses to companies for oil and gas exploration, three of which are located in the so far unexplored areas in the Barents Sea close to the Arctic (Government of Norway 2016). According to some studies, the proposed new fields for oil and gas drilling would result in GHG emissions 150% higher than from the existing fields (Oil Change International 2017). Norway also has a favourable taxation regime, which allows companies to recover most of their costs invested in oil and gas exploration. The esti-mates show that Norway has subsidized oil and gas companies with €9.7 billion within a decade through this favourable tax regime (Bellona 2017).

In sum, Norway has demonstrated a strong willingness to disrupt car-bon-based investments internationally, particularly coal that may compete with Norwegian gas exports. This policy has supported Norway´s efforts to maintain visibility in the international arena and shape global climate governance. The corporatist structures with a strategic role in the oil and gas industry at home has meant that the systemic support for the domestic fossil fuel industry remained untouched.

Strategic goals for constraining carbon-based technologies

Early Norwegian climate policy, adopted in 1989, focused on stabilizing national CO_2 emissions at 1989 levels by the year 2000 (Hovden and Lindseth 2002). This approach was soon abandoned and replaced by flex-ible international mechanisms as the most cost-effective instrument for reducing GHG emissions. Under the first Kyoto Protocol period, Norway committed not to exceed its 1990 level of GHG emissions by more than 1% through 2012. One should note the active role played by the Norwegian government in shaping the design of the Kyoto Protocol and ensuring full respect for the principles of cost-effectiveness and flexibility. The govern-ment voluntarily proposed increasing its target to 10% reduction by 2012, although Norway failed to achieve that goal, as its GHG emissions in 2012 were 4.5% higher than in 1990. It met the voluntary Kyoto target of 10% reduction only in 2015 (prior to the Paris Climate Summit) by purchasing international emissions credits. Under the second Kyoto Protocol period, Norway took on the target of 16% reduction by 2020 compared to 1990, pledging to meet two-thirds of this through domestic reductions. Norway continued to agree voluntary to increasingly ambitious climate mitigation targets despite its very slow progress in meeting previous commitments. In 2012, Norway further raised its target to 30% emissions reductions by 2020 and to achieve carbon-neutrality by 2050 (Norwegian Ministry of Environment 2012). In 2015, a tentative goal of at least 40% reductions by 2030 (with the EU) was put forward (Norwegian Ministry of Climate and Environment 2015). In the transport sector, the government declared the goal of 85g GHG emissions per km on average for all new cars by 2020, which is 10% higher than the corresponding EU target (Figenbaum et al. 2013). Furthermore, according to the new National Transport Plan for 2018–2029, all new passenger cars and light vans should be zero-emission by 2025 (Norwegian Ministry of Transport and Communications 2017). The ambitious climate goals, an important part of Norway´s foreign policy strategy, have not been fully met given the domestic political-economic challenges. Despite considerable efforts, Norwegian GHG emissions in 2016 were 3.3% higher compared to the baseline year 1990 (Statistics Norway 2017).

Discussion

On the basis of our analysis we can distinguish three main phases in Norway's climate policy mix considering its creative and destructive dimensions. The first phase (1989–1995), termed 'symbolic destruction', emphasised cost-effective stabilization of domestic emissions, with a modest carbon tax as the main policy instrument. In the second phase (mid-1990s–2008) attention shifted towards incremental improvements in environmental efficiency, with the focus on cost-effective international flexibility mechanisms. Creative destruction remained modest and niche support weak. The third phase (since 2008) has been characterized by increasing efforts towards emissions-reductions at home combined with intensive global climate diplomacy and multiple initiatives aimed at promoting low-carbon niche technologies and disrupting fossil-fuel investments internationally. During this phase, the goal of industrial upgrading and exports based on low-carbon sectors and technologies has gained prominence. Existing instruments (CO_2 tax and electric vehicle subsidies) have been strengthened and new policy instruments added, such as market incentives for established renewable-energy technologies and R&D spending on selected new low-carbon technologies. International creative and disruptive climate policy efforts have been reinforced and diversified. However, the actual impact of these cumulative changes on climate policy has been limited, and has hardly affected the prevailing ´policy equilibrium´ (Cashore and Howlett 2007) based on the cautious creation and destruction policy efforts at home and high innovativeness and activity abroad.

We have sought to demonstrate how the mixed efforts for mitigating climate change domestically and high activity in the global climate governance regime have been tied to Norway´s political economy and foreign policy strategy as a small, open, social-investment model of economy. Regarding creative destruction, democratic corporatism and inherited political pragmatism and cost-effectiveness combined with the exceptional economic clout of the petroleum sector, have ensured that control policies remained modest and systemic support for R&D and a favourable business climate for the oil and gas industry continued unchallenged. Driven by its foreign policy concerns to maintain and shape global climate agreements, Norway has launched multiple ´disruptive´ international initiatives such as divestment from coal in its Pension Fund and the measures for reforming fossil-fuel subsidies globally. These have served to strengthen Norway´s norm-advocacy in international climate diplomacy and ease the pressure for changes at home. On the niche creation side, Norway has hesitantly adopted the binding EU renewable energy target, which failed to create the dynamic domestic market for renewable energy with most projects implemented in Sweden through the green certificate scheme. Less-mature and more costly technologies such as offshore wind have found even less support from policy-makers and established industries. The political and economic consensus for supporting niche low-carbon energy technologies has been present for CCS due to its attractiveness for the incumbent fossil-fuel industries. In the transport sector, the lack of strong vested interests and high taxation policy on vehicles have played a facilitating role in enabling effective niche creation for electric cars. The lead market for electric cars has contributed to mitigating domestic GHG emissions but has also empowered Norway's norm-setter role in international climate governance. In addition to generally modest domestic niche creation efforts, Norway has increased its global financial commitment to low-carbon technologies through developmental aid, EU funds and the Green Climate Fund.

Overall, these insights support the theoretical notion that small corporatist economies are successful in incremental long-term adaptation but prone to ignore larger structural problems. Whereas Norway has achieved a remarkably stable climate policy consensus and has continuously encouraged environmental improvements in the oil and gas sector, it has failed to formulate a plan for phasing out oil and gas extraction and reduce the country's economic dependence on oil and gas exports. Another important finding is the influential role of Norwegian foreign policy for national climate policymaking. As noted by Cass (2008), climate policy provides fertile ground for symbolic politics as many governments feel obliged to abide by international climate norms but are often not willing or forced to act upon international commitments with tangible and costly policy reforms. Although Norway's climate policy can partly be described as symbolic, for instance the modest CO_2 tax, international climate policy pledges have motivated Norway to undertake some more structural reforms such as the extensive promotion of electric vehicles. This indicates that small countries with social-investment economies such as Norway may be more inclined to scale-up their climate policy efforts in the face of external pressure as they are more dependent on stable international agreements and a progressive self-image than are larger and more liberal-market oriented economies such as Australia and Canada (see Cass 2008). The extent to which international climate agreements will induce domestic policy change depends not only on the level of ambition of the goals but also on the implementation mechanisms. The existing international and EU climate governance regimes offer considerable flexibility for countries such as Norway to fulfil their climate commitments without engaging in deeper emission cuts and creative destruction at home.

Conclusions

Here, we aimed to offer an updated and innovative analysis of the creative and disruptive character of Norway's climate policy mix, linking it to the theory-based concept of small states. We might highlight at least two major contributions that we believe we have made to the literatures on sustainable transition and climate policy and politics.

First, we have demonstrated the value of the 'niche support vs. creative destruction' categorization proposed by Kivimaa and Kern (2016) by providing a more nuanced understanding of the national climate policy mixes, their scope and impact. We have further developed the proposed framework by offering a more consistent way of categorizing key creative and destructive policy measures. We have also included Norway's climate mitigation efforts abroad in the analytical framework, which adds to the comprehensiveness and robustness of the analysis. The application of this framework has revealed important insights into Norway's climate policy mix, its strong international character, and particularly its failure to disrupt and constrain the domestic oil and gas industry as the major producer and exporter of GHG emissions. Important differences in the creative and destructive policy impacts have been detected between the energy and transportation sectors, which suggests the importance of sectoral differences and cross-sectoral comparisons in studying climate policies.

Second, we have made the case for linking the national political-economic setting and foreign policy strategy to explain the character and dynamics of climate policies. The structural political-economic and foreign policy features of Norway as a small advanced type of corporatist-investment economy together with some specific attributes of Norway's political-economic setting and foreign policy identity have greatly influenced the pace and design of national climate policy, measured by niche creation and creative destruction policy measures. Overall, our analysis has demonstrated the value of conceptualizing different state models based on their material and ideational base to facilitate better understanding and comparison of the creative and disruptive potential of national climate policy mixes.

Disclosure statement

No potential conflict of interest was reported by the authors.

References

Bellona, 2017. Bellona tries to pull the plug on vast Norwegian oil and gas exploration subsidies. Available from: https://bellona.org/news/fossil-fuels/2017-08-23814

Björkdahl, A., 2008. Norm advocacy: a small state strategy to influence the EU. *Journal of European Public Policy*, 15 (1), 135–154. doi:10.1080/13501760701702272

Blindheim, B., 2015. A missing link? The case of Norway and Sweden: does increased renewable energy production impact domestic greenhouse gas emissions? *Energy Policy*, 77, 207–215. doi:10.1016/j.enpol.2014.10.019

Brende, B., 2015. *Foreign policy address to the Storting*. Available from: http://www.norway.org/News_and_events/Embassy/FM-Brendes-Foreign-Policy-Address-March-5-2015/#.Vz-rlY9OKM9

Carrington, D., 2015. *Norway confirms $900bn sovereign wealth fund's major coal divestment*. Guardian. Available from: http://www.theguardian.com/environment/2015/jun/05/norways-pension-fund-to-divest-8bn-from-coal-a-new-analysis-shows

Cashore, B. and Howlett, M., 2007. Punctuating which equilibrium? Understanding thermostatic policy dynamics in Pacific Northwest forestry. *American Journal of Political Science*, 51 (3), 532–551. doi:10.1111/j.1540-5907.2007.00266.x

Cass, L.R., 2008. A climate of obstinacy: symbolic politics in Australian and Canadian policy. *Cambridge Review of International Affairs*, 21 (4), 465–482. doi:10.1080/09557570802452763

Ćetković, S., *et al.* 2017. Varieties of clean energy transitions in Europe: political-economic foundations of onshore and offshore wind development. *In*: D. Arendt, ed. *The political economy of clean energy transitions*. Oxford: Oxford University, 102–123.

David, M., 2017. Moving beyond the heuristic of creative destruction: targeting exnovation with policy mixes for energy transitions. *Energy Research & Social Science*, 33, 138–146. doi:10.1016/j.erss.2017.09.023

Dryzek, J.S., 2003. *Green states and social movements: environmentalism in the United States, United Kingdom, Germany, and Norway*. New York: Oxford University Press.

Dyrstad, J.M., 2015. *Resource curse avoidance: governmental intervention and wage formation in the Norwegian petroleum sector*. Department of Economics Working Paper Series No.6/2015, Norwegian University of Science and Technology.

Eckersley, R., 2016. National identities, international roles, and the legitimisation of climate leadership: Germany and Norway compared. *Environmental Politics*, 25 (1), 180–201. doi:10.1080/09644016.2015.1076278

EEA Grants, n.d. *How the grants are spent*. Available from: http://eeagrants.org/Results/How-the-Grants-are-spent

ENERGI21, 2014. *National strategy for research, development and commercialisation of new energy technology*.

ENOVA, 2015. *Results and activities 2015*. Available from: http://viewer.zmags.com/publication/9513c0bc#/9513c0bc/1

ERKC, n.d. Norway. Available from: https://setis.ec.europa.eu/energy-research/country/norway

European Commission, 2017. *Trade, countries and regions: Norway*. Available from: http://ec.europa.eu/trade/policy/countries-and-regions/countries/norway/

Export Credit Norway, 2014. *Annual report 2014*.

Fæhn, T., *et al.* 2013. *Climate policies in a fossil-fuel producing countries: demand versus supply side policies*. Discussion paper 747, Statistics Norway.

Fagerberg, J., Mowery, D., and Verspagen, B., eds, 2009. *Innovation, path dependency and policy: the Norwegian case*. Oxford: Oxford University Press.

Falleti, T.G., 2016. Process tracing of extensive and intensive processes. *New Political Economy*, 21 (5), 455–462. doi:10.1080/13563467.2015.1135550

Figenbaum, et al. 2013. 85g pr kilometer i 2010. Er det mulig? TØI rapport 1264/ 2013. Oslo: Institute of Transport Economics.

Geels, F.W., *et al.*, 2017. Sociotechnical transitions for deep decarbonization. *Science*, 357 (6357), 1242–1244. doi:10.1126/science.aao3760

Globalcarbonproject, 2017. *Global co2 emissions likely to rise in 2017.* Available from: http://www.globalcarbonproject.org/carbonbudget/17/files/Norway_ CICERO_GCPBudget2017.pdf

Government of Norway, 2008. *Agreement on Norway's climate policy.* Available from: https://www.regjeringen.no/globalassets/upload/md/vedlegg/klima/agree ment_on_norways_climate_policy_080117.pdf

Government of Norway, 2014a. *Norges strategi for internasjonalt samarbeid for reform av subsidier til fossile brensler.* Available from: https://www.regjeringen. no/no/dokumenter/strategi-fossile-brensler/id765000/

Government of Norway, 2014b. *The agreement on climate policy.* Available from: https://www.regjeringen.no/en/topics/climate-and-environment/climate/innsikt sartikler-klima/agreement-on-climate-policy/id2076645/

Government of Norway, 2015. *Norway to scale up support for the green climate fund.* Available from: https://www.regjeringen.no/en/aktuelt/green_fund/ id2465040/

Government of Norway, 2016. *Announcement 23rd licensing round awards.* Available from: https://www.regjeringen.no/en/aktuelt/announcement-23rd-licensing-round-awards/id2500936/

GSI and IISD, 2012. *Fossil fuels – at what cost? Government support for upstream oil and gas activities in Norway.* Geneva, Switzerland: International Institute for Sustainable Development, Global Subsidies Initiative.

Gvalia, G., *et al.*, 2013. Thinking outside the bloc: explaining the foreign policies of small states. *Security Studies*, 22 (1), 98–131. doi:10.1080/09636412.2013.757463

Hall, P.A. and Soskice, D.W., 2001. *Varieties of capitalism: the institutional foundations of comparative advantage.* Oxford: Oxford University Press.

Hansen, M.E. and Gjefsen, T., 2015. *The end of Nordic exceptionalism.* Available from: https://www.kirkensnodhjelp.no/globalassets/utviklingskonf-2015/end-of-nordic-exceptionalism.pdf

Harris, P.G., 2002. Bringing the in-between back in: foreign policy in global environmental politics. *Politics & Policy*, 36 (6), 914–943. doi:10.1111/j.1747-1346.2008.00145.x

Hofmann, B., Jevnaker, T., and Thaler, P., 2019. Following, challenging, or shaping: Can third countries influence EU energy policy? Politics and Governance, 7 (1), 152–164. doi:10.17645/pag.v7i1.1853

Hovden, E., *et al.* 2002. Norwegian climate policy 1989–2002. *In*: W.M. Lafferty, ed. *Realizing Rio in Norway: evaluative studies of sustainable development.* Oslo: ProSus, 143–168.

IEA, 2017. *IEA find co2 emissions flat for third straight year even as global economcy grew in 2016.* Available from: https://www.iea.org/newsroom/news/2017/march/ iea-finds-co2-emissions-flat-for-third-straight-year-even-as-global-economy-grew.html

IFE, 2015. *Energy efficiency trends and policies in Norway.* Kjeller: Institute for Energy Technology.

Ingebritsen, C., *et al.*, 2006. *Small states in International relations.* Reykjavik: University of Iceland Press.

Ingebritsen, C., 2010. Katzenstein's legacy 25 years after: small states in world markets. *European Political Studies*, 9 (3), 359–364. doi:10.1057/eps.2010.23

Innovation Norway, n.d. *Innovation Norway gets half a billion Norwegian kroner to the green shift.* Available from: http://www.innovasjonnorge.no/en/start-page/our-services/sustainability/innovation-norway-gets-half-a-billion-norwegian-kroner-to-the-green-shift/

Katzenstein, P.J., 1985. *Small states in world markets: industrial policy in Europe.* Ithaca, NY: Cornell University Press.

Keating, M., 2015. The political economy of small states in Europe. *In*: H. Baldersheim and M. Keating, eds. *Small states in the modern world: vulnerabilities and opportunities.* Cheltenham: Edward Elgar, 3–23.

Keating, M. and Harvey, M., 2014. The political economy of small European states: and lessons for Scotland. *National Institute Economic Review*, 227, 54–66. doi:10.1177/002795011422700107

Kivimaa, P. and Kern, F., 2016. Creative destruction or mere niche support? Innovation policy mixes for sustainability transitions. *Research Policy*, 45 (1), 205–217. doi:10.1016/j.respol.2015.09.008

Lachapelle, E. and Paterson, M., 2013. Drivers of national climate policy. *Climate Policy*, 13 (5), 547–571. doi:10.1080/14693062.2013.811333

Mjøset, L. and Cappelen, Å., 2011. The integration of the Norwegian economy into the world economy. *Comparative Social Research*, 28, 167–263.

Neumann, I.B. and Gstöhl, S., 2004. *Lilliputians in Gulliver's world: small states in international relations.* Reykjavik: Centre for Small State Studies, University of Iceland.

Normann, H.E., 2014. *Offshore wind and carbon capture and storage in Norway.* Presentation, TIK – Centre for Technology, Innovation and Culture, University of Oslo.

Normann, H.E., 2015. The role of politics in sustainable transitions: the rise and decline of offshore wind in Norway. *Environmental Innovation and Societal Transitions*, 15, 180–193.

Norwegian Ministry of Climate and Environment, 2015. *New emission commitment for Norway for 2030 – towards joint fulfilment with the EU.* Meld. St. 13 (2014–2015) Report to the Storting (white paper). Available from: https://www.regjeringen.no/contentassets/07eab77cc38f4085abb594a87aa19f10/en-gb/pdfs/stm201420150013000engpdfs.pdf

Norwegian Ministry of Climate and Environment, 2017. *Norway´s electric vehicles policy.* Available from: http://www.mhsr.sk/uploads/files/ocrMJ55J.pdf

Norwegian Ministry of Foreign Affairs, 2009. *Interests, responsibilities and opportunities: the main features of Norwegian foreign policy.* Report No.15 (2008–2009) to the Storting.

Norwegian Ministry of Petroleum and Energy, 2017. Agreement on Swedish expansion of the certificate scheme. https://www.regjeringen.no/en/aktuelt/agreement-on-swedish-expansion-of-the-certificate-system/id2549653/

Norwegian Ministry of the Environment, 2012. *Norwegian climate policy.* Report No. 21 (2011–2012) to the Storting (white paper), summary. Available from: https://www.regjeringen.no/contentassets/aa70cfe177d2433192570893d72b117a/en-gb/pdfs/stm201120120021000en_pdfs.pdf

Norwegian Ministry of Transport and Communications, 2017. National transport plan 2018–2029, Report to the Storting (white paper). Available from: https://www.regjeringen.no/contentassets/7c52fd2938ca42209e4286fe86bb28bd/en-gb/pdfs/stm201620170033000engpdfs.pdf

Norwegian Petroleum, 2019. The government's revenues. https://www.norskpetroleum.no/en/economy/governments-revenues/

NVE & Energimyndigheten, 2016. *The Norwegian-Swedish electricity market: annual report 2016*. Available from: https://energimyndigheten.a-w2m.se/Home.mvc?ResourceId=5676

OECD, 2016. *OECD.Stat: level of GDP per capita and productivity*. Available from: https://stats.oecd.org/Index.aspx?DataSetCode=PDB_LV

Oil Change International, 2017. *The sky´s limit Norway. Why Norway should lead the way in a managed decline of oil and gas extraction*. Washington: Oil Change International.

Pauly, L.W. and Jentleson, B.W., 2014. *Power in a complex global system*. London and New York: Routledge.

Peters, B.G., Jordan, A., and Tosun, J., 2017. Over-reaction and under-reaction in climate policy: an institutional analysis. *Journal of Environmental Policy & Planning*, 19 (6), 612–624. doi:10.1080/1523908X.2017.1348225

Research Council of Norway, 2015. *Det norske forskningsog innovasjonssystemet – statistikk og indikatorer*. Lysaker: Norges forskningsråd.

Roettereng, J.-K.S., 2016. How the global and national levels interrelate in climate policymaking: foreign policy analysis and the case of carbon capture storage in Norway's foreign policy. *Energy Policy*, 97, 475–484. doi:10.1016/j.enpol.2016.08.003

Rogge, K.S. and Johnstone, P., 2017. Exploring the role of phase-out policies for low-carbon energy transitions: the case of the German energiewende. *Energy Research & Social Science*, 33, 128–137. doi:10.1016/j.erss.2017.10.004

Rogge, K.S. and Reichardt, K., 2016. Policy mixes for sustainability transitions: an extended concept and framework for analysis. *Research Policy*, 45 (8), 1620–1635. doi:10.1016/j.respol.2016.04.004

Schoenefeld, J.J., Hildén, M., and Jordan, A.J., 2018. The challenges of monitoring national climate policy: learning lessons from the EU. *Climate Policy*, 18 (1), 118–128. doi:10.1080/14693062.2016.1248887

Skånland, Ø.H., 2010. 'Norway is a peace nation': a discourse analytic reading of the Norwegian peace engagement. *Cooperation & Conflict*, 45 (1), 34–54. doi:10.1177/0010836709347212

Skjærseth, J.B. and Christiansen, A.C., 2006. Environmental policy instruments and technological change in the energy sector: findings from comparative empirical research. *Energy & Environment*, 17 (2), 223–241. doi:10.1260/095830506777070024

Statistics Norway, 2017. *Emissions of greenhouse gases*. Available from: https://www.ssb.no/en/klimagassn

Statkraft, 2014. *Billions for renewable energy*. Available from: http://www.statkraft.com/media/news/News-archive/2014/Billions-for-renewable-energy/

Statkraft, 2015. No new investments in offshore wind. Available from: https://www.statkraft.com/media/news/News-archive/20151/no-new-investments-in-offshore-wind/

Divergent neighbors: corporatism and climate policy networks in Finland and Sweden

Antti Gronowⓘ , Tuomas Ylä-Anttilaⓘ , Marcus Carson
and Christofer Edlingⓘ

ABSTRACT
Previous research has suggested that corporatist polities tend to enact more ambitious environmental policies than others. Here it is argued that the macro concept of corporatism can be dissected into three components: inclusiveness, consensualism and strength of tripartite organisations. These components of corporatism can be measured at the meso-level of policy networks. It is proposed that inclusiveness and consensualism are related to ambitious climate policy but exclusive tripartite coalitions can be detrimental for the ambitiousness of climate policy. This argument is backed by evidence from policy network surveys in two similar corporatist countries where climate change policies diverge: Sweden, where policies are ambitious, and Finland, where they are less so. It is found that in Sweden the climate change policy network is more consensual and slightly more inclusive, while in Finland tripartite organisations play a strong role.

Introduction

It is frequently argued that corporatism is generally good for the environment. Corporatist polities, it is claimed, tend to enact more ambitious environmental policies than others (for example, Christoff and Eckersley 2011). However, divergent voices argue that at least in some cases, corporatism may hinder rather than promote ambitious environmental policy (Dryzek *et al.* 2002).

We suggest that these contradictory arguments can be explained by a tendency in the literature to oversimplify corporatism, treating it either as a dichotomous macro structural variable – a category that a country either fits or not – or a continuum on which all countries can be placed depending on the degree of corporatism. Instead, we suggest that corporatism can be divided into three components: inclusiveness, consensualism and strength of tripartite organisations, and that these components are visible at the meso-level of policy networks. While inclusiveness and consensualism are indeed conducive to ambitious environmental policy, tripartite strength seems to have the opposite effect. Thus, by focusing on the meso-level of policy networks, our analysis takes the debate about corporatism and environmental policy beyond the sometimes contradictory findings of earlier macro comparative research. Our results show that variation not only between but also within political institutions can be significant. Based on this contribution, we suggest that 'corporatism' is too abstract a macro level variable to provide insight into the mechanisms of environmental policy outcomes.

We use a classical 'most similar systems' research design (Przeworski and Teune 1970), comparing Finland and Sweden, two similar countries that diverge in the ambitiousness of their climate policies. Both countries are examples of the Nordic corporatist polity type, and their natural environment and social organisation are in many ways very similar. But Finland's per capita emissions are double those of Sweden, and climate policy output in Sweden is clearly more ambitious. Thus, while the two countries are similar in their macro-level social structures, we look for differences in their meso-level policy networks. Our main finding, as our analytical framework leads us to expect, is that in Sweden the climate change policy network is more inclusive and slightly more consensual, while in Finland tripartite organisations play a stronger role.

Our contribution is organised as follows. We begin by reviewing the literature on corporatism and environmental policy, drawing on the literature to formulate our analytical framework and hypotheses. We then present the case countries, summarising their similarities and also how their climate change policies differ. We then present the data, our network analytical methods, and results, before concluding with a discussion of the implications of our findings.

Analytical framework and hypotheses

Most literature on the relationship between corporatism and environmental policy argues that corporatist polities are pioneers of environmental policy-making (for reviews of relevant literature, see Fiorino 2011, Koch and Fritz 2014). For example, Liefferink et al. (2009) found that corporatist institutional structures are favourable to the advancement of environmental regulation. Wälti (2004) showed that corporatist countries perform better in reducing NO_2 and SO_2 pollution. Scruggs (2003) argued that countries characterised by 'neo-corporatist institutions have enjoyed better environmental performance than countries where economic groups are less comprehensively organised and policy making is less consensual' (Scruggs 2003, p. 123).

Arguments explaining why corporatist countries fare better in environmental policy refer to characteristics often associated with the idea of corporatism, including broad participation in decision making, dialogue, and a consensual style of policy making. Jacob and Volkery (2006, p. 84) argued that the environmental performance of corporatist countries is due to the 'integrative capacity of a country, i.e. a style of governance that does not restrict participation to general elections but organises it in a broad and consensual way.' Jänicke (2005) came to similar conclusions in his framework for trendsetters in environmental policy. He sees a policy style centred on dialogue and consensus as one of the conditions for successful environmental policy.

Christoff and Eckersley (2011) argued that the organisational structure of corporatist polities ensures that a small group of players can consensually work towards a common goal. Thus, implementation of policies that serve broader interests and overcome problems of collective action is made possible (see also Matthews 2001). Theories of ecological modernisation assume that ecological protection and business interests are not necessarily irreconcilable. It is possible that corporatist institutions facilitate ecological modernisation by bringing together diverse actors, thereby fostering a common understanding of the importance of environmental issues (cf. Zannakis 2009). Furthermore, it is sometimes suggested that corporatism reduces veto points where powerful interest groups can block climate policy initiatives (Karapin 2016, p. 54).

There are, however, divergent voices. Koch and Fritz (2014) found that corporatist social-democratic welfare regimes are no better than liberal ones in introducing green dimensions into their policies. Karapin (2012) argued that corporatist institutions may even be harmful for the environment if they institutionalise the power of producer groups who favour economic growth over environmental protection. Dryzek *et al.* (2002) go so far as to argue that the Nordic variant of corporatism (they only analyse Norway and generalise from this) is bad for the environment because it tends to lead to co-optation of environmental non-governmental organisations (NGOs) by the government.

As noted above, we argue that these contradictory assessments of the relationship of corporatism and environmental protection can result from treating corporatism as a unitary, macro-structural variable (where a country is either corporatist or not) or as a continuum on which all countries can be placed according to their corporatist characteristics (Lijphart 2012). Such analyses lump together all features of corporatism without considering the possibility that some of these features may strengthen environmental policy performance while others hinder it. We argue that it is useful to distinguish, in particular, between consensualism, inclusiveness and strength of tripartite organisations.

Consensualism, inclusiveness and tripartite strength are variables that can be difficult to measure at the macro-level. We propose to measure consensualism, inclusiveness and tripartite strength as features of meso-level social organisation. This means that while these characteristics might be difficult to measure as countries' macro properties, they are reflected and visible in how policy networks in the climate change policy domain are organised. Following Jepperson and Meyer (2011), when attempts at macro-level explanations seem too abstract to provide convincing answers, it makes sense to look for explanations at the meso-level of policy networks. Our objective here is therefore to add a new dimension to the mostly macro-comparative literature on corporatism and environmental policy by using meso-level concepts, methods, and materials from the field of policy network research.

Below, we draw four hypotheses concerning the differences of climate change policy networks in Finland and Sweden, combining ideas from the literature on corporatism and environmental policy with concepts from policy network analysis. The hypotheses concern the manifestations of the three components of corporatism we have outlined above – inclusiveness, consensualism and tripartite strength – at the policy network level. First, corporatist polities are *inclusive*. In this polity type, NGOs and the state are closely linked. The Nordic model, in particular, is characterised by the state traditionally being tightly integrated with civil society (Rothstein and Trägårdh 2007). The political opportunity structure is usually very open in the Nordic model (Ylä-Anttila 2010).

At the level of policy networks, we expect corporatist inclusiveness to manifest in relatively equal power distribution in the network (cf. Knoke 1994, Schneider 2015) and relatively equal access of different organisational sectors to governmental decision-making processes. Equal access provides no guarantee that power is distributed equally, but makes it more likely. Indeed, a central feature of corporatist polities is that they do not concentrate all power in the hands of the executive. Furthermore, it is likely that a balance of power between environmental organisations, business and labour interests would prevent economic policy objectives from overriding ecological ones and encourage responses aiming to reconcile both perspectives. If this assumption holds true, the more ambitious climate policy of Sweden would be associated with the Swedish policy network being more inclusive than the Finnish network. Our inclusiveness hypothesis, then, states:

H1: The Swedish policy network is more inclusive than the Finnish policy network regarding the distribution of influence between organisations and access to decision-making processes.

Second, corporatist polities are consensual. The inclusive nature of the political system, the self-understanding centred on closeness between state and civil society, and the extensive consultation processes organised to support policymaking supposedly translate, in the end, into a broad-based consensus on the course of action to be taken (Alapuro 2005, Ylä-Anttila 2010, Luhtakallio 2012). Indeed, consensualism is the feature of corporatism most often-mentioned by those who argue that corporatism strengthens environmental policymaking. For example, Jacob and Volkery (2006) argue that corporatism entails an integrative style of governance. If it is true that consensualism is good for ambitious environmental policy, the more ambitious country – Sweden – should be more consensual. Our consensualism hypothesis, then, states that:

H2: There is a stronger consensus on climate change policy in the Swedish policy network than in the Finnish policy network.

Third, in corporatist polities there is a strong tripartite bargaining system made up of trade unions, business peak organisations, and state organisations. This bargaining system extends beyond collective bargaining over wages into areas such as environmental policy. Karapin (2012) argues that this characteristic of corporatism may not be good for the environment. Echoing the classic Treadmill of Production theory (Schnaiberg and Gould 1994), he observes that corporatist institutions may be harmful for environmental quality if producer groups' power is institutionalised and favours an economic growth agenda at the expense of environmental concerns (Karapin 2012, p. 48; see also Hukkinen 1995). If the strength of tripartite organisations is indeed a hindrance to ambitious environmental performance, we expect to find these organisations to be stronger in the less well performing of the two countries – Finland. Thus, our tripartite strength hypothesis states:

H3: Industry and labour peak organisations are more influential in the Finnish than in the Swedish policy network.

Finally, it may be that the strength of tripartite organisations is a function of their alliances with other organisations in the policy network. One of the most prominent theoretical approaches within the policy network literature, the Advocacy Coalition Framework (ACF), argues that it is not the influence of individual organisations that is crucial in determining policy output but, rather, coalitions of like-minded, collaborating organisations. From this perspective, differences in policy between two countries follow from differences in the opinions, resources and network positions of competing advocacy coalitions (Sabatier 1998, Kriesi and Jegen 2001, p. 252).[1]

In the case of climate change policymaking, this would mean that the tripartite organisations prioritising economic growth over the environment are not only influential, but also able to assemble coalitions that include influential allies. Such coalitions might include governmental organisations and political parties, and these connections to the state are what give tripartite organisations their power. If it is true that the strength of network coalitions led by tripartite organisations is an important hindrance to ambitious environmental policy, we should expect to find stronger coalitions of this type in the less ambitious of the two countries we study – Finland. Thus, our coalition hypothesis states:

H4: The advocacy coalitions involving industry and labour peak organisations are more influential and better connected to the state in the Finnish network than in the Swedish network.

Case countries

In many ways, our case countries, Finland and Sweden, are similar: small Nordic states with cold climates, advanced postindustrial economies, and consensual corporatist political systems. Lijphart's (2012) ranking of countries, which is based on their interest group pluralism versus corporatism, places Finland the fifth and Sweden the most corporatist country of the 36 countries compared. The political institutions found in both countries help define Nordic corporatism, which is characterised by tripartite agreements, strong peak organisations and multi-party polities (Lane and Ersson 2002). In Esping-Andersen's (1990) well-known classification of welfare regimes, both countries are classified as social-democratic. Besides embodying corporatist institutions, Finland and Sweden are both small countries with relatively open economies.

In other words, from a macro-comparative perspective, both formal institutional structures and political cultures are fairly similar. This institutional similarity is partly due to a long common history–until 1809 Finland was a part of the Kingdom of Sweden. The two countries' natural and human environments are also similar. Cold climates require considerable energy for heating and this is equally true for both countries. Low population density requires much energy for transport, and this also characterises both countries almost equally: Finland has 18 inhabitants per square kilometer; Sweden has 24 (compared, for instance, to Germany at 232 and South Korea at 518). The greater share of both countries' populations resides in the south.

Finland and Sweden's greenhouse gas (GHG) emissions differ considerably, however. Finland's per capita emissions have been around double the level of Swedish emissions for the past two decades. Sweden's per capita emissions have declined steadily since 1990 (reference year for the Kyoto Protocol) as shown in Figure 1. Finland's emissions have recently dropped as well, but largely due to a severe economic recession. Finland's gross domestic product (GDP) fell dramatically by 8.3% in 2008, with negative or zero growth rates persisting through 2012–2015 (Statistics Finland 2017a). The drop in the Swedish GDP was less severe, (−0.7% in 2008 and −5.1% in 2009), and the rebound was quick: the economy grew 3.8% in 2015 (OECD 2017). Thus, the reduction in Finnish emissions seen in Figure 1 between 2012 and 2015 was probably a result of a shrinking economy, while in Sweden emissions were reduced despite strong economic growth. Finland's per capita emissions are still almost double those of Sweden. With the Finnish economy recovering (in 2017 GDP rose by 2.8%), Finland's emission levels are on the rise again.[2]

Important structural influences on emissions are built into the respective countries' energy infrastructures and economic base, and while these are strongly influenced by policy choices, they involve sunk costs that can result in long time lags for realising CO_2 reductions. Three key factors affect the extent to which policy has influenced the energy mix, and thereby GHG emissions: energy production and use; the types of domestically available energy sources and degree to which each country depends on them; underlying factors contributing to 'energy intensity', which very roughly speaking, is an inverse measure of energy efficiency (IEA 2014).

For example, according to the International Energy Agency's Energy System Overview (2016), over half (57%) of Sweden's 155 TWh of electricity was generated using renewables, with hydropower contributing 40% of

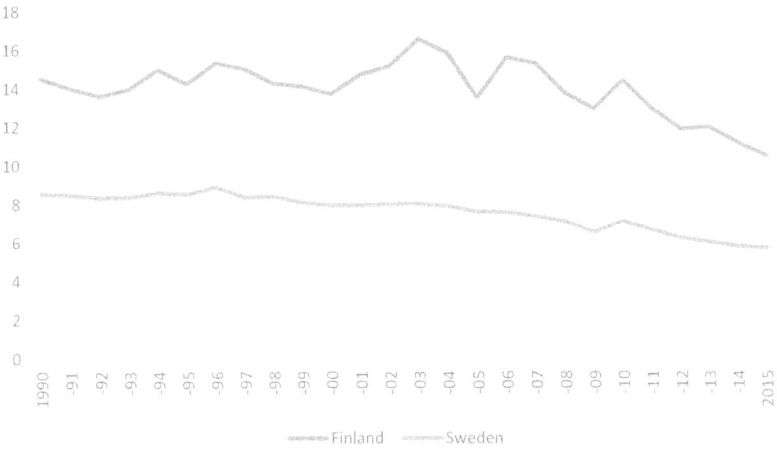

Figure 1. GHG emissions (metric tons of CO_2 equivalent), per capita in Finland and Sweden 1990–2015.

Data source: European Environment Agency (2017).

the total. Nuclear provided most of the rest (41%), with oil, gas and coal accounting for a tiny 2.4%. Finland's electricity production, totaling less than half of Sweden's at 68.8 TWh, was generated with renewables accounting for 45%, a far smaller share for hydropower (21%), and a larger share produced by biofuels (10%). Much of the difference in emissions can be traced to Finland's continued use of coal and oil for electricity generation. As in Sweden, oil and gas generate very little of Finland's electricity, but coal generation accounts for 11%.

What is especially notable is that neither country has any meaningful domestic coal/gas/oil reserves. This means that both countries' fossil fuel supplies must be imported, contributing little to energy security. Sweden imports 42% of its crude oil from the Russian Federation (and 30% from neighbouring Norway), and all of its natural gas from Denmark. Over half of Sweden's modest coal imports come from Australia (54%). Finland imports 88% of its crude oil, all of its natural gas, and 64% of its coal from the Russian Federation.

An important practical implication of the absence of domestic fossil fuel production is that the kind of internal influence of fossil fuel producers on policy found elsewhere, such as in the United States (where coal currently accounts for over 40% of electricity production), is absent – i.e. not part of domestic policy networks. Economic actors that use coal would, however, be part of those networks. The economic structure of the two countries could therefore help explain policy differences and emissions trajectories. For example, Karapin (2016, p. 56) has argued that a large service sector in relation to manufacturing can be 'beneficial for climate policy since service-sector firms have lower costs of adjustment and are less exposed to international competition than manufacturing firms'.

Here again, the profiles of the countries are fairly similar. Both Swedish and Finnish economies are export-based and differences in the proportion of manufacturing and services have been relatively minor in recent years. In 2016, the GDP share of all industrial activity was 27% in both countries, while services accounted for 71% in Finland and 73% in Sweden (Statista 2018a, 2018b). Differences in these proportions have diminished over time and while it is possible that future industrial activity could account for an increasing share as Finland recovers from its recession, the differences would still remain relatively minor. It is unlikely that the Swedish economy being slightly more service-based would explain the difference in performance in energy and climate policy. GHG emissions from fuel combustion in the transport sector have fallen in Sweden (from 21.1 in million tons of CO_2 equivalent in 2005 to 17.9 in 2014). The figures are lower in Finland and also seem to be declining (from 13.0 in million tons of CO_2 equivalent in 2005 to 11.1 in 2014). However, Sweden's population is almost twice as large and, instead of a clear declining trend, Finland's figures appear to be fluctuating (see Statista 2018c, 2018d).

The Finnish economy is also consistently more energy intensive (see Figure 2), although energy intensity dropped slightly after 2010 due to an economic recession that hit the export industry particularly hard. Moreover, Finland's domestic material consumption is more dependent on fossil energy materials. According to Eurostat, in 2015 Finland and Sweden consumed 3.5 and 1.7 tonnes of fossil energy materials per capita respectively. The large differences in energy intensity between the two countries can probably be explained more by fuel mix and Sweden's more decisive move away from fossil fuels than by the nature of economic activity.

It has been suggested that the price of electricity is seen in Finland as an issue of national competitiveness and that this concern has tended to outweigh climate change mitigation as an issue (Teräväinen 2012). However, electricity prices are similar to those in Sweden, especially for non-household consumers (Eurostat 2017). With the price of electricity no lower in Finland than in Sweden, it could still be that keeping electricity prices down is seen as a higher priority in Finland.

Beyond simple emission trajectories, Sweden is considered the more ambitious country of the two on climate policy. The Germanwatch Climate Change Performance Index, which evaluates climate policy performance, has always placed Sweden at the top, whereas Finland consistently

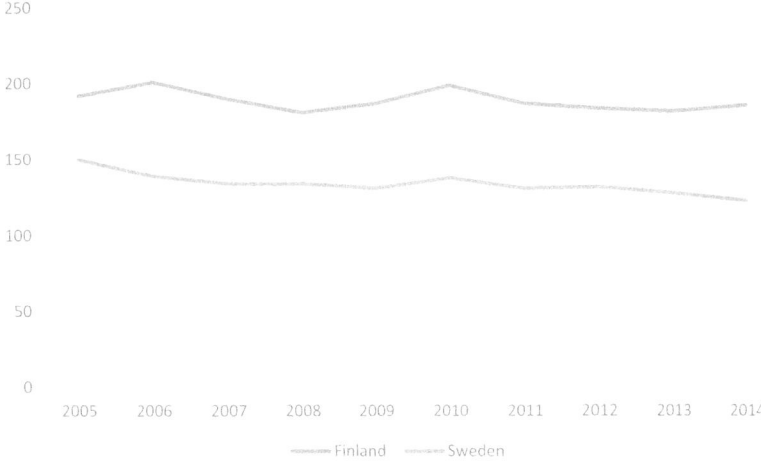

Figure 2. Energy intensity of the Finnish and Swedish economies, 2005–14.
Kg of oil equivalent per 1 000 EUR of GDP.
Data source: Eurostat (2016): Energy, transport and environment indicators.

does much worse (Germanwatch 2017). On a ten-year average (2008–2017), Sweden was the second best performer in the world, while Finland was 31st.

In the process of defining the EU's total GHG reduction commitments and the burden-sharing within the EU, Sweden has been a forerunner, while Finland has often dragged its feet (Tirkkonen 2000, Teräväinen 2012). When preparing for the Kyoto Protocol, arguments around the economy and national interest dominated in Finland, with industry organisations especially claiming that proposed Kyoto targets presented a major threat to economic growth and national competitiveness (Tirkkonen 2000, p. 117). Finland was eventually persuaded by international pressure to accept the Kyoto and EU targets, and both countries were among the first in the world to implement national carbon taxes. In Finland, however, energy-intensive industries soon managed to secure exemptions from these taxes (Tamminen *et al.* 2016).

In 2017 Sweden set itself a target of completely phasing out its GHG emissions by 2045 (The Independent 2017). This target, more ambitious than in any other developed nation, is part of a law that was drawn up by a cross-party committee. The current Finnish Minister of Environment has proposed that Finland should also aim to become carbon neutral by 2045. However, this is only a non-binding proposal and it assumes that emission levels might be balanced (rather than phased out) by using forests as carbon sinks (Valtioneuvosto 2017).

The environmental policy literature consistently lists Sweden as a pioneer in environmental issues (Sarasini 2009) and one of the few countries implementing effective climate policies (Karapin 2016, p. 2). Yet Finland has been labeled 'a failing eco-state' due to its ecological performance, particularly in the policy domain of climate change (Koch and Fritz 2014). Overall, the data on emissions and policy outputs, and expert academic assessments all point to a clear difference in the levels of ambition between the climate policies of the two countries.

Data and methods

Our data is derived from network surveys conducted in 2014 in Finland and in 2015 in Sweden. The data were collected as a part of the Comparing Climate Change Policy Networks research project (see compon.org). The respondents are representatives of all major organisations interested in climate policy in these countries. The response rate was 83% for Finland and 66% for Sweden, with 146 responses in all.

A preliminary list of respondent organisations was compiled based on previous research and knowledge of climate policy in each country. It was ensured that organisations from different sectors of society (e.g., business, government, and NGOs) would be represented and that the respondent organisations would be fairly similar in both countries. The original lists of respondents were presented to experts on climate policy who suggested some additions and omissions. The final list included 96 organisations in Finland and 104 organisations in Sweden. The respondents were first contacted by phone and if they agreed to participate, a link to an online questionnaire was sent by email.

Our measures of the three different features of corporatism, derived from the survey data, are as follows.

Inclusiveness

To test hypothesis 1, our first measure of the inclusiveness of the policy networks is the distribution of influence among the organisations belonging to them. Our questionnaire included a roster of all organisations in the network, and respondents were asked to check off organisations they considered influential in climate change policymaking. The influence score produced for each organisation was normalised to a scale ranging from 0 to 1. The power structure perspective suggests that the distribution of influential positions in a policy network is a measure of the concentration of power in a political system (Knoke 1994, Schneider 2015), and more concentration equals less inclusiveness. We measure the concentration or the dispersion of power with the Lorenz curve, which is usually used by economists to measure levels of (in)equality of wealth or income. In this case, we are thus measuring the concentration of power based on our respondents' assessments of influence (cf. Schneider 2015).

Our second measure of inclusiveness is access to policymaking processes. The measure is self-reported, based on the answer to the question 'How often does your organisation take part in policymaking, such as formal testimony at hearings, participation in government advisory committees, or drafting legislation proposals or text?', measured on a scale ranging from 1 to 3 (never/sometimes/often). The more equally this access is distributed between different organisation types, the more inclusive the policy network should be. For example, if all NGOs lacked access, we would interpret this finding as an indication of exclusiveness.

Consensualism

To test hypothesis 2, our measure of consensualism is based on a composite variable measuring pro-climate beliefs of the organisations in the policy network. Here, 'pro-climate' refers to indicating that climate change policy should be prioritised on a level comparable to economic growth. The composite variable consists of six items measuring pro-climate beliefs. Based on inter-item correlation tables and a series of exploratory factor analyses of 21 belief variables, we identified six strongly correlated items that constituted the final scale. The items ranged from the validity of climate science to desirability of governmental mitigation efforts and prioritisation of mitigation over economic growth. We calculated the simple sum of all items (as opposed to summing by factor scores) to maximise transparency of interpretation and scaled the final composite variable so that values ranged from 0 to 1 to maximize readability. Cronbach's alpha for the composite variable is 0.875 for Finland and 0.828 for Sweden (see Appendix 1). The higher the standard deviation of this variable in the network, the less consensual we take the network to be, because high standard deviation indicates lack of agreement.

Tripartite strength

To test hypothesis 3, our first measure of the strength of tripartite collective bargaining organisations is the influence scores outlined above (see inclusiveness). The higher the mean score for business peak organisations and trade unions, the stronger the influence of tripartite organisations in the policy network. Second, we compare the influence of tripartite organisations and NGOs to investigate the proposition that it may not be the absolute strength of tripartite organisations but, rather, their strength in relation to environmental organisations that leads to poor environmental policy performance. Thus, we define the *influence gap between NGOs and tripartite organisations* as the difference between the mean influence of NGOs and of tripartite organisations in the policy network. Because both items are measured on a scale from 0 to 1, the influence gap can range between –1 (all-powerful NGOs and tripartite organisations with no power) and +1 (all-powerful tripartite organisations and NGOs with no power).

Finally, to test hypothesis 4, we analyse the coalition structures of the policy networks in both countries. Our survey included a question about who the respondents collaborate with on climate policy at the organisational level. Respondents were presented with rosters that listed all the respondent organisations of the surveys – that is, the major organisations in both countries having a stake in climate policy issues. From these lists, respondents were asked to indicate organisations they regularly collaborate with.

The collaboration networks were symmetrised using the maximum criterion, which makes all ties reciprocal. Thus, a collaborative tie exists if it is reported by ego and/or alter.[3] This symmetrisation enables us to include non-respondent nodes in the network; otherwise these nodes would have to be excluded because network analysis requires that matrices are square. In order to find coalitions, the collaboration network in both countries was split into subgroups using the factions routine in UCInet, which fits the nodes into subgroups that have highest possible internal density. The number of subgroups is decided by assessing the *proportion of correctness* that shows which solution has the highest goodness of fit. The proportion of correctness is a measure of the total number of 'errors,' which refers to absent within-faction and present between-faction ties. The factions routine needs to be run several times in order to see whether the factions are stable or whether there are nodes that could plausibly be placed in two or more factions. There were some isolates in the Swedish data, and these were excluded from the analysis of subgroups because isolates are by definition outsiders in relation to any subgroupings.

In addition to the collaboration relationships, we also measure the internal belief homophily of coalitions by looking at the standard deviations of the above-mentioned composite scale on climate change beliefs within coalitions.

Results

Inclusiveness

Our first hypothesis postulated that the Swedish policy network is more inclusive than the Finnish policy network regarding the distribution of influence and access to policymaking. The Lorenz curves, which measure the distribution of influence in both countries, are shown in Figure 3.

The curves in Figure 3 show the distribution of influence as cumulative percentages. In a completely equal policy network, the distributions would follow the equality line. The figure shows curves differing from the equality line in both policy networks, which is unsurprising because even a consensual polity is unlikely to distribute power completely equally. We postulated that power in the Finnish network would be more concentrated than in Sweden. However, the curves follow almost identical trajectories, showing that concentration is similar in both cases. Therefore, regarding the distribution of influence, our inclusiveness hypothesis is rejected because the policy networks in both countries are equally inclusive.

Our second inclusiveness measure, which compares access to policymaking, displays differences per country. We compare the level of access that trade unions, business peak organisations, and NGOs have on a scale of 1 to 3. We are not reporting figures for governmental actors for political access because governmental organisations should have access by definition. Moreover, we are mainly interested in seeing how NGOs manage compared to tripartite actors. Table 1 shows that NGOs fare relatively well in both countries. Finland's score is actually higher for NGOs than Sweden's (2.25 compared to 2.00). More interesting, however, is that in Sweden NGOs score higher than either trade unions or business peaks, whereas in Finland, NGOs score lower than business peaks or trade unions. Thus, Swedish NGOs seem to have better access into official policymaking

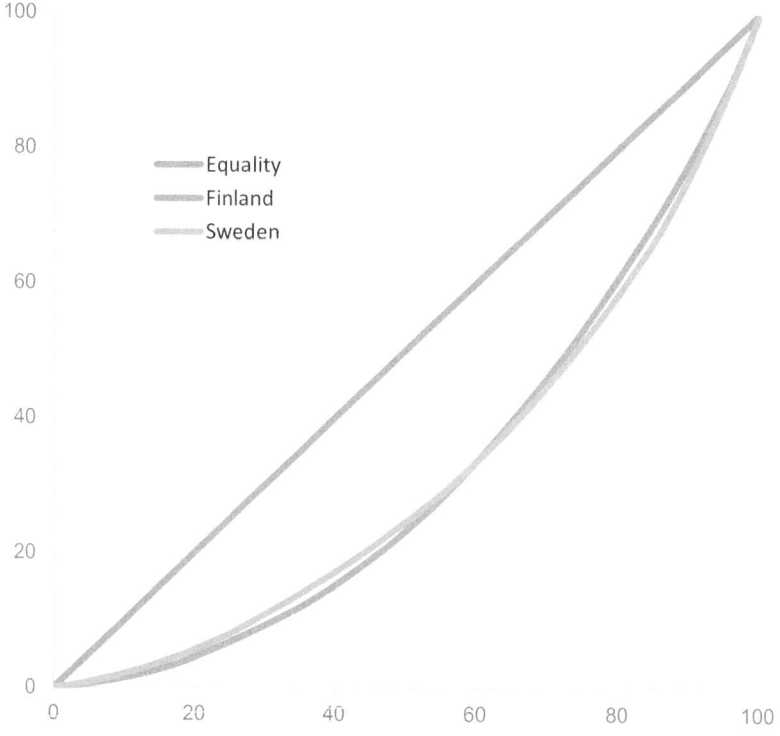

Figure 3. The Lorenz curves for influence in Finland and Sweden.

than either trade unions or business peak organisations, with the reverse true in Finland.

Overall, the evidence concerning inclusion seems inconclusive because there are no differences in the distribution of influence between Finland and Sweden, but NGOs are more integrated into official policymaking in Sweden than in Finland. However, this fact concerning integration only holds if one compares NGOs with unions and business peaks.[4]

Consensualism

Our second hypothesis postulated that the Swedish policy network would be more consensual. First, analysis of the pro-mitigation beliefs of networks in both countries shows that the Swedish network is slightly more pro-climate. On a scale from 0 to 1, where higher figures indicate pro-mitigation beliefs, the mean score of the Swedish network is 0.79, while in Finland the corresponding figure is 0.75. Focusing on the extent of consensus reveals that Sweden is more consensual, but only slightly. Standard deviation of pro-climate beliefs is 0.17 in Sweden and 0.21 in Finland.

However, when we zoom in on the consensualism of pro-climate beliefs among the 15 most influential organisations in each country's policy network, the results change somewhat. These organisations are more supportive of pro-mitigation policies in Sweden (mean 0.73) than in Finland (0.64). Furthermore, consensus on pro-mitigation beliefs among them – measured by standard deviations of the composite variable – is clearly higher in Sweden (0.15) than in Finland (0.29) (the closer to zero the standard deviation, the more consensual the network). Thus, the relatively minor differences between Finland and Sweden in both pro-mitigation beliefs and the extent of consensus become more pronounced when examining the 15 most influential organisations. Overall, our second hypothesis is not fully rejected since differences between the two countries exist.

Tripartite strength

Our third hypothesis posited that tripartite collective bargaining organisations are more influential in the Finnish policy network than the Swedish policy network. This does indeed seem to be the case, as Table 1 illustrates. Mean influence scores for trade unions and business peak organisations are higher than for NGOs in Finland. In Sweden, NGOs are slightly less influential than business peaks but more influential than unions. Especially notable is the score of business peaks in Finland: it is highest of all organisational sectors. In Sweden, business peaks are also influential, but the difference compared with NGOs is nowhere near as high.

Turning next to the influence gap between tripartite organisations and NGOs, this measure can range between –1 (all-powerful NGOs, no tripartite organisations power) and +1 (all-powerful tripartite organisations, no NGO power). The result shows a positive figure for Finland and negative for Sweden (see Table 2). Thus, Finnish tripartite actors are more powerful than NGOs, with the opposite true in Sweden, which aligns with our hypothesis.

Table 1. Inclusiveness and influence in Finland and Sweden.

	Inclusiveness		Influence	
	Finland	Sweden	Finland	Sweden
Trade Unions	2.75	1.80	0.27	0.11
Business Peak Organisations	2.75	1.67	0.42	0.25
NGOs	2.25	2.00	0.18	0.20

Note: Inclusiveness measures show how often an organization takes part in policymaking on a scale from 1 (never) to 3 (often). Influence measures the perceived influence of an organisation on a standardized scale from 0 to 1.

Table 2. Summary of main results.

Characteristics of corporatism and their measures	Sweden	Finland
Inclusiveness	**High**	**High**
Concentration of power	Relatively equal	Relatively equal
Access to decision making of NGOs and tripartite organisations	NGOs more dominant	Others more dominant
Consensualism	**High**	**Lower**
Pro-climate beliefs	High (0.79)	Slightly lower (0.75)
Consensus on pro-climate beliefs	High (STD 0.17)	Slightly lower (STD 0.21)
Pro-climate beliefs, top 15 influential organisations	High (0.73)	Lower (0.64)
Consensus on pro-climate beliefs, top 15 influential organisations	High (STD 0.15)	Lower (STD 0.29)
Tripartite Strength	**Lower**	**High**
Influence of tripartite organisations	Lower (0.18)	High (0.35)
Influence gap between NGOs and tripartite organisations	Negative (−0.02)	Positive (0.17)
Advocacy coalitions	None	Tripartite, Government, NGO
Ties of tripartite coalition to the state	N.A.	Strong

Testing our fourth hypothesis through network analysis provides more support for this argument. Hypothesis 4 posited that advocacy coalitions involving tripartite organisations are more influential and better connected to the state in the Finnish network than in the Swedish network. The first step in our coalition analysis finds that the Finnish collaboration network is more clustered than Sweden's. Overall densities indicate that in Finland, 26% of all possible ties are present, compared to 11% in Sweden. The weighted overall clustering coefficients for Finland and Sweden were 0.43 and 0.27, respectively, indicating that the Finnish network is more clustered into collaborating subgroups. To establish the number of coalitions, we make the ideal-typical assumption that the network consists of a given set of clusters that are internally dense, but which have no external connections. The factions algorithm fits the nodes into subgroups based on internal density.

The best-fitting solution for Finland was one with four subgroups. The proportion of correctness of this solution was relatively high, at 0.74. We find three subgroups with high within-density and a group of outsiders that are not connected with each other (and thus do not constitute a coalition). However, the factions are not stable which means that there are organisations whose placement keeps changing with reruns of the algorithm. Many of these are peripheral in the network but some could be considered brokers that broker relations between actual coalitions. For further analyses, we removed these organizations and the faction of outsiders and were left with a core of 53 organisations.[5] We think these actors constitute actual coalitions because they mainly collaborate with each other.

In line with the Advocacy Coalition Framework (ACF), the coalitions are constituted not only by collaboration ties but also by similarity of beliefs, as measured by our pro-mitigation beliefs composite variable. Based on the constituency of the three subgroups and the policy core beliefs of these organisations, we judge that these three subgroups are indeed coalitions (see also Gronow & Ylä-Anttila 2016). We label them Tripartite, Government and Research, and NGO Coalition. Their within-densities are high: 0.66 (Tripartite), 0.68 (Government and Research), and 0.65 (NGO). All of these densities are much higher than the average density of the whole network (0.26). The coalitions therefore exhibit dense collaboration ties.

The differences in the beliefs of the three coalitions are statistically significant (one-way ANOVA F = 14,36, p = 0,000). The Tripartite Coalition is the least ecologically minded (mean 0.57), the NGO coalition the most (0.97). The Government and Research Coalition falls between them (0.74). In Finland, the most influential organisations are not only less supportive of ambitious climate change policy, but are also members of the Tripartite Coalition. Interestingly, both from the point of view of the ACF theory and of policy outcomes, this coalition is no mere business lobby group. Instead, it groups together business peak organisations, trade unions, and two big political parties (right and centre). This coalition is the most influential: 0.42 compared to 0.15 of the NGO coalition as measured by reputational power on a scale from 0 to 1. In addition, the Tripartite Coalition is also well connected to the government. This can be seen in its connections to the Government and Research Coalition: it has more collaboration relationships across coalition lines (inter-coalition density 0.24) than the Environment Coalition (inter-coalition density 0.12).

For Sweden, it is much more difficult to find subgroups that could be considered coalitions. A two-faction solution is relatively stable; most of the organisations are delegated to only one faction with several reruns of the algorithm. However, the proportion of correctness of this solution is very low at 0.56. In addition, ACF argues that the coalitions are also constituted by similarity of beliefs and there are almost no differences in the pro-mitigation beliefs between the two factions (measured by our composite variable). Models that add more factions (we experimented with up to eight factions) produce better results regarding proportion of correctness but they are very unstable. Theoretically it makes little sense to argue that there are eight coalitions with distinctly different beliefs about climate change politics. Indeed, ACF maintains that there are usually two to three coalitions in a policy domain. Overall, we conclude that the lack of belief congruence within factions and stable solutions indicates that there seem to be no proper coalitions in the Swedish policy network.

Summarising our results (see Table 2), we find that the concentration of power is almost identical in both countries and that Swedish NGOs fare better than their Finnish colleagues in comparison with trade unions and business peaks. We also find that the Swedish network is slightly more pro-mitigation than its Finnish counterpart. Moreover, Sweden is also more consensual in beliefs concerning climate change mitigation, but this is a close call. However, the 15 most influential organisations in Sweden are more pro-mitigation, with consensus on these issues more pronounced. Thus, general consensualism is not higher in the Swedish network, but consensualism at the level of influential actors can make a difference. Finally, we find that in Finland, NGOs are less influential than tripartite organisations, whereas in Sweden the reverse is true (the so called influence gap). Only in Finland can we find subgroups or coalitions and also a coalition of which both tripartite organisations are part. Thus, in Finland the tripartite organisations are influential and occupy important positions in a resourceful coalition well linked to the government. This contrasts with Sweden and appears as an important difference in the climate policy networks between the two countries.

Conclusion

We set out to scrutinise the claim that corporatist polities tend to enact more ambitious environmental policies than others. Using a similar systems research design, we compared Finland and Sweden, two similar corporatist polities where climate change policy differs – Sweden is clearly the more ambitious of the two. Instead of treating 'the corporatist polity type' as a macro variable, the prevalence of which can be measured by a single index in a country, we differentiated between three key aspects of corporatism: inclusiveness, consensualism, and the strength of tripartite collective bargaining organisations. We have focused on each of these issues at the meso-structural level of policy networks in order to find the ways in which each aspect is related to climate policy.

Our findings indicate that one can identify particular features of corporatism at the level of policy networks that act as an elaboration of the connection between corporatism and environmental policy – climate policy in this case. Even if the distribution of power in the network is roughly equal, assessing the political access of NGOs compared with tripartite actors shows different trajectories. Furthermore, consensualism of the whole network can be too crude a measure, but if many influential actors are in favor of ambitious climate policy and the network in general more or less agrees on this point, then more ambitious policy is a probable result, as the Swedish case shows. Furthermore, we suspect that tripartite connectivity can be an obstacle for such policy if tripartite actors end up in the same political coalition, which excludes NGOs.

A shortcoming of our quantitative policy network approach is the relatively scant attention it pays to the role of meaning. We have analysed pro-mitigation beliefs of the policy networks but there are other, historical reasons for the observed policy differences having to do with meaning-making. For instance, Sweden's highly international and pioneering role in climate science probably has an influence on the importance that policy actors ascribe to climate change (Zannakis 2009).

One could make the objection that our results simply reflect a more energy intensive Finnish economy. Even if this were the case, our results still show meso-level factors to which corporatism gives rise in combination with an energy-intensive economy. Furthermore, we suspect that once certain meso-level structural features are in place (e.g. an influential tripartite coalition), they exert a semi-independent causal force on future policy. Longitudinal data collection is a prerequisite for unveiling such dynamics, and the current cross-sectional design is a weakness of our analysis. By disentangling the time dynamics, a longitudinal design would allow firmer conclusions about causes and effects to be drawn. However, like the present study, most policy network research is cross-sectional due to difficulties of obtaining longitudinal policy network data.

Nevertheless, we have shown that variation within political institutions (e.g. corporatism) can be as important as variation between institutions when it comes to climate policy. This makes it worthwhile to analyse the nature of corporatism in greater detail – as we have done. One of the implications of our findings for future research is the replication of the research setting in other corporatist countries. It would be especially interesting to see what kind of dynamics of consensualism, inclusiveness and tripartite strength can be found in non-Nordic corporatist countries (e.g. Switzerland).

Notes

1. Policy network analysis is often more interested in the exchange of resources than the Advocacy Coalition Framework (ACF) which sees the role of policy-related beliefs as essential. However, we tend to think of the ACF and policy network analysis as complementary perspectives.
2. GHG emissions grew in Finland by as much as 6% in 2016 (Statistics Finland 2017b). Swedish emission levels were also on the rise, growing around 3.5%, perhaps as a sign of continued economic growth (Statistics Sweden 2017). This is still much less than in Finland (with similar GDP growth rates).
3. In Finland, two respondents reported that they collaborate with almost all the 96 listed organisations. These responses deviated so much from the others (and for the most part their ties were not reciprocated by alters) that these organisations' ties were symmetrised using the minimum criterion, which includes only reciprocal ties.
4. The numbers of organisations are small in each category, so one should be cautious when interpreting these comparisons.

5. As a robustness check, we also performed all analyses without removing the possible broker organizations. We did this by randomly assigning the brokers to the nearest factions. This does not change the main results. However, running the analyses with the smaller core group tends to amplify the differences between coalitions (e.g. differences in policy core beliefs). We take this result to indicate that the organizations belonging to the core group are indeed the ones that represent actual coalitions.

Disclosure statement

No potential conflict of interest was reported by the authors.

Funding

Funding for this research was provided by the Academy of Finland (Grant No. 266685 and No. 298819), the Kone Foundation (Grant No. 085319), and the Swedish Research Council (Grant No. 2007-2363).

ORCID

Antti Gronow (iD) http://orcid.org/0000-0002-7639-6422
Tuomas Ylä-Anttila (iD) http://orcid.org/0000-0002-6908-3495
Christofer Edling (iD) http://orcid.org/0000-0002-3909-1080

References

Alapuro, R., 2005. Associations and contention in France and Finland: constructing the society and describing the society. *Scandinavian Political Studies*, (28), 377–399. doi:10.1111/j.1467-9477.2005.00137.x

Christoff, P. and Eckersley, R., 2011. Comparing state responses. *In*: J.S. Dryzek, R. B. Norgaard, and D. Scholsberg, eds. *The Oxford handbook of climate change and society*. Oxford: Oxford University Press, 431–448.

Dryzek, J.S., *et al.*, 2002. Environmental transformation of the state: the USA, Norway, Germany, and the UK. *Political Studies*, 50 (4), 659–682. doi:10.1111/1467-9248.00001

Esping-Andersen, G., 1990. *The three worlds of welfare capitalism*. Cambridge: Polity Press.

Eurostat, 2016. *Energy, transport and environment indicators*. http://ec.europa.eu/eurostat/web/products-statistical-books/-/KS-DK-16-001 (accessed 1 September 2017

Eurostat, 2017. *Electricity price statistics*. http://ec.europa.eu/eurostat/statistics-explained/index.php/Electricity_price_statistics (accessed 10 January 2018

Fiorino, D.J., 2011. Explaining national environmental performance: approaches, evidence, and implications. *Policy Sciences*, 44 (4), 367–389. doi:10.1007/s11077-011-9140-8

Germanwatch, 2017. *The Climate Performance Index 2017*. https://germanwatch.org/en/13042 (accessed 29 August 2017).

Gronow, A. and Ylä-Anttila, T., 2016. Cooptation of ENGOs or treadmill of production? Advocacy coalitions and climate change policy in Finland. *Policy Studies Journal.* doi:10.1111/psj.12185

Hukkinen, J., 1995. Corporatism as an impediment to ecological sustenance: the case of Finnish waste management. *Ecological Economics*, 15 (1), 59–75. doi:10.1016/0921-8009(95)00019-6

IEA, 2014. *World energy outlook 2014.* International Energy Agency. https://www.iea.org/newsroom/news/2014/november/world-energy-outlook-2014.html (accessed 10 January 2018

The Independent, 2017. *Sweden pledges to reach net-zero carbon emissions by 2045.* 17 June. http://www.independent.co.uk/news/sweden-greenhouse-emissions-latest-2045-parliament-climate-act-donald-trump-paris-agreement-a7794686.html (accessed 25 August 2017

Jacob, K. and Volkery, A., FFU Report 01/2006, 2006. Modelling capacities for environmental policy-making in global environmental politics. *In:* M. Jänicke and K. Jacob, eds. *Environmental governance in global perspective.* Freie Universität Berlin, 67–94.

Jänicke, M., 2005. Trend setters in environmental policy. *European Environment*, 15 (2), 129–142. doi:10.1002/eet.375

Jepperson, R. and Meyer, J.W., 2011. Multiple levels of analysis and the limitations of methodological individualisms. *Sociological Theory*, 29 (1), 54–73. doi:10.1111/j.1467-9558.2010.01387.x

Karapin, R., 2012. Explaining success and failure in climate policies: developing theory through German case studies. *Comparative Politics*, 45 (1), 46–68. doi:10.5129/001041512802822879

Karapin, R., 2016. *Political opportunities for climate policy: California, New York, and the federal government.* New York: Cambridge University Press.

Knoke, D., 1994. *Political networks: the structural perspective.* Cambridge: Cambridge University Press.

Koch, M. and Fritz, M., 2014. Building the eco-social state: do welfare regimes matter? *Journal of Social Policy*, 43 (4), 679–703. doi:10.1017/S004727941400035X

Kriesi, H. and Jegen, M., 2001. The Swiss energy policy elite: the actor constellation of a policy domain in transition. *European Journal of Political Research*, 39 (2), 251–287. doi:10.1111/ejpr.2001.39.issue-2

Lane, J.-E. and Ersson, S., 2002. The Nordic Countries. Contention, compromise and corporatism. *In:* J.M. Colomer, ed. *Political institutions in Europe.* 2nd ed. London & New York: Routledge, 245–278.

Liefferink, D., *et al.*, 2009. Leaders and laggards in environmental policy: a quantitative analysis of domestic policy outputs. *Journal of European Public Policy*, 16 (5), 677–700. doi:10.1080/13501760902983283

Lijphart, A., 2012. *Patterns of democracy. Government forms and performance in thirty-six countries.* 2nd. New Haven & London: Yale University Press.

Luhtakallio, E., 2012. *Practicing democracy: local activism and politics in France and Finland.* Eastbourn: Palgrave-MacMillan.

Matthews, M., 2001. Cleaning up their acts: shifts of environment and energy policies in pluralist and corporatist states. *Policy Studies Journal*, 29 (3), 478–498. doi:10.1111/j.1541-0072.2001.tb02105.x

OECD, 2017. *Economic Outlook: Statistics and Projections.* https://data.oecd.org/gdp/real-gdp-forecast.htm (accessed 30 August 2017).

Przeworski, A. and Teune, H., 1970. *The logic of comparative social inquiry.* New York: Wiley.

Rothstein, B. and Trägårdh, L., 2007. The state and civil society in a historical perspective: the Swedish case. *In*: L. Trägårdh, ed. *State and civil society in Northern Europe. The Swedish model reconsidered.* New York and Oxford: Berghahn, 229–253.

Sabatier, P.A., 1998. The Advocacy coalition framework: revisions and relevance for Europe. *Journal of European Public Policy*, 5 (1), 98–130. doi:10.1080/13501768880000051

Sarasini, S., 2009. Constituting leadership via policy: Sweden as a pioneer of climate change mitigation. *Mitigation and Adaptation Strategies for Global Change*, 14 (7), 635–653. doi:10.1007/s11027-009-9188-3

Schnaiberg, A. and Gould, K.A., 1994. *Environment and society: the enduring conflict.* New York: St. Martin's.

Schneider, V., 2015. Towards post-democracy or complex power sharing? Environmental policy networks in Germany. *In*: V. Schneider and B. Eberlein, eds. *Complex democracy. Varieties, crises, and transformations.* Chem: Springer, 263–279. doi:10.1007/978-3-319-15850-1_17

Scruggs, L., 2003. *Sustaining abundance: environmental performance in industrial democracies.* Cambridge: Cambridge University Press.

Statista, 2018a. Finland: share of economic sectors in the gross domestic product (GDP) from 2006 to 2016. *The Statistics Portal.* https://www.statista.com/statistics/327513/share-of-economic-sectors-in-the-gdp-in-finland/(accessed   10 January 2018)

Statista, 2018b. Sweden: share of economic sectors in the gross domestic product (GDP) from 2006 to 2016. *The Statistics Portal.* https://www.statista.com/statistics/375611/sweden-gdp-distribution-across-economic-sectors/(accessed   10 January 2018)

Statista, 2018c. Annual greenhouse gas emissions from fuel combustion in the transport sector in Finland from 2005 to 2014 (in million tons of CO_2 equivalent). *The Statistics Portal.* https://www.statista.com/statistics/411914/annual-greenhouse-gas-emissions-of-the-transport-sector-in-finland/(accessed 12 January 2018)

Statista, 2018d. Annual greenhouse gas emissions from fuel combustion in the transport sector in Sweden from 2005 to 2014 (in million tons of CO_2 equivalent). *The Statistics Portal.* https://www.statista.com/statistics/411915/annual-greenhouse-gas-emissions-of-the-transport-sector-in-sweden/(accessed 12 January 2018)

Statistics Finland, 2017a. *National accounts.* http://tilastokeskus.fi/til/vtp/2016///vtp_2016_2017-07-13_tie_001_en.html (accessed 30 August 2017

Statistics Finland, 2017b. *Suomen kasvihuonekaasupäästöt 2016.* https://www.stat.fi/til/khki/2016/khki_2016_2017-05-24_kat_001_fi.html (accessed 15 January 2018)

Statistics Sweden, 2017. *Greenhouse gas emissions continue to increase.* http://www.scb.se/en/finding-statistics/statistics-by-subject-area/environment/environmental-accounts-and-sustainable-development/system-of-environmental-and-economic-accounts/pong/statistical-news/environmental-accounts–emissions-to-air-q3-2016/(accessed 15 January 2018)

Tamminen, S., Ollikka, K., and Laukkanen, M., 2016. Suomen energiaverotus suosii energiaintensiivisiä suuryrityksiä. VATT Policy Brief 2-2016. http://vatt.fi/suomen-energiaverotus-suosii-energiaintensiivisia-suuryrityksia (accessed 8 January 2017)

Teräväinen, T., 2012. *The politics of energy technologies. Debating climate change, energy policy, and technology in Finland, the United Kingdom, and France.* Helsinki: Into.

Tirkkonen, J., 2000. *Ilmastopolitiikka ja ekologinen modernisaatio. Diskursiivinen tarkastelu suomalaisesta ilmastopolitiikasta ja sen yhteydestä metsäsektorin muutokseen.* [Climate policy and ecologial modernisation. A discursive study of Finnish climate policy and its connections with forestry.]. Acta Universitatis Tamperensis, 781. Tampere: Tampere University Press

Valtioneuvosto, 2017. *Minister Tiilikainen: Finland to achieve carbon-neutrality by 2045.* http://valtioneuvosto.fi/artikkeli/-/asset_publisher/ministeri-tiilikainen-suomesta-hiilineutraali-yhteiskunta-viimeistaan-vuonna-2045?_101_INSTANCE_3wyslLo1Z0ni_languageId=en_US (accessed 20 September 2017).

Wälti, S., 2004. How multilevel structures affect environmental policy in industrialized countries. *European Journal of Political Research*, 43 (4), 599–634. doi:10.1111/j.1475-6765.2004.00167.x

Ylä-Anttila, T., 2010. *Politiikan paluu. Globalisaatioliike ja julkisuus.* [The Return of Politics. The Global Justice Movement and the Public Sphere.]. Tampere: Vastapaino.

Zannakis, M., 2009. *Climate policy as a window of opportunity. Sweden and global climate change.* Gothenburg Studies in Politics 121. Gothenburg: University of Gothenburg.

Appendix 1. The composite variable measuring pro-mitigation beliefs[a,b,c].

Survey item	Loading (FI)	Communality (FI)	Loading (SE)	Communality (SE)
Climate change is not caused by human activities	0.742	0.551	0.709	0.502
Climate science is too uncertain to be a basis for policymaking	0.736	0.542	0.578	0.334
National economic competitiveness is more important than taking care of climate change	0.822	0.676	0.826	0.682
Securing national energy supply is more important than taking care of climate change	0.825	0.681	0.772	0.596
The government puts too much effort into reducing CO_2 emissions	0.843	0.710	0.727	0.528
My country should not try to take a leading international role in international negotiations on climate change	0.789	0.623	0.783	0.614

[a]Extraction method: principal component analysis.
[b]All items load onto one component explaining 63.06% of total variance in Finland and 57,27% in Sweden.
[c]Cronbach's alpha for composite variable is 0.875 for Finland and 0.828 for Sweden.

The politics of carbon taxation: how varieties of policy style matter

Mikael Skou Andersen

ABSTRACT

The momentum achieved for unilateral carbon taxes in seven European countries is examined. Why is it that small countries, despite being vulnerable to forces of international competition, have been able to implement carbon taxes? A review of national experiences does not suggest that the share of fossil fuels in the energy mix defines the room for such taxes, or point to a strong role for traditional left-right ideology. Rather, it is deep-seated patterns of national policy styles with neo-corporatist traits, providing a protective device for the open economies of small countries, which condition the introduction of carbon taxes. The associated routines of decision-making offer coordination mechanisms for proactive macro-economic policies in which carbon taxation can find a place. Parliamentary democracies with proportional representation, as is common in the smaller countries, provide access to government for political parties that pursue carbon taxation. These in turn sensitise larger political parties to climate concerns, as they benefit from institutionalised practices and routines for problem-solving and consensus-seeking.

Introduction

The idea of taxing carbon emissions materialised almost 30 years ago at the International Conference of the Changing Atmosphere in Toronto, the first to address both the science and policy of climate change (Toronto Conference 1988). The emerging scientific consensus astonished its participants, who called for immediate action to curb emissions, including a 'levy on fossil fuel consumption in industrialised countries' as a measure to raise funds to support developing countries.

With an established tradition of taxing energy products other than motor fuels, it was not difficult for four Nordic countries to include a carbon component in their excise taxation.[1] Well before the 1992 Rio declaration, Finland pioneered the very first carbon tax (1990). Sweden (1991), Norway (1991) and Denmark (1992) followed soon after. Adoption was by no means uncontroversial and concerns over competitiveness were voiced by industry and labour unions. Nevertheless, by acting as climate mitigation pioneers it was expected that other countries would follow their example, creating a more level playing field.

Unfortunately, these hopes proved to be wishful thinking. The Netherlands (1992) had indeed introduced a carbon tax component under its pollution tax scheme, but it was soon abandoned in favour of a small users' energy tax (Andersen 1996). Despite an EU leadership role in global climate negotiations, none of the medium-size or larger European countries was willing unilaterally to adopt carbon taxes, although they edged close to doing so on several occasions. Germany, for instance, announced a carbon tax in its 1990 reunification treaty, while Italy even passed the required legislation (1999), only to suspend it soon after (OECD 2002, p. 138). The UK enacted a Climate Change Levy (2001), but with a tax base related to energy rather than carbon, to safeguard domestic coal mining, and earning the reputation of an 'eclectic' tax (Pearce 2006, p. 155).

Initially, Nordic countries retrenched. Sweden, for example, lowered the rate of its carbon tax (Speck 2009). However, from the late 1990s, all four Nordic countries expanded their carbon tax bases and tax rates, thus enabling a reduction of their relatively high labour and payroll taxes. These tax-shifting processes materialised despite the shelving of the European Commission's carbon tax proposal (Zito 2002). Gradually, several smaller countries have adopted carbon taxes, most recently Portugal (2014), while Iceland (2013), Ireland (2010), Switzerland (2008), Croatia (2007), Estonia (2000) and Slovenia (1997) have also found ways to place taxes on carbon. There are, however, often special arrangements for business and energy-intensive industries, as referred to below. France (2014), acting in anticipation of the Paris meeting, remains the exception among larger European countries in adopting a carbon tax (Sénit 2012).[2]

These observations provide the starting point for the research question addressed here: how and why is it that small countries, whose emissions are virtually irrelevant to global climate change and which are more open to forces of international competition, are more likely to unilaterally introduce carbon taxes? The hypothesis is that variations in national policy styles of regulation help explain such policy outputs.

The significance and varieties of national policy styles

Climate change policy relies on a broad range of policy instruments, many of which have been agreed within the EU framework, reflecting EU requirements for unanimity, but also some that remain the prerogative of the nation state. Carbon taxation, although believed to be critical to a timely curbing of emissions, belongs to the latter category, involving most of the controversial issues related to unilateral climate policy action. EU Member

States carry with them different perceptions of the appropriate relationship between state and market. These differences relate not only to political ideologies and beliefs, but are far deeper and rooted in historical experiences and approaches, having caused different varieties of national policy styles to emerge.

These varieties were highlighted in the 'capitalism-versus-capitalism' debate that followed the collapse of planned economies in Europe. Following Albert's (1993, 1996) terminology, a Manchester approach to capitalism favours a less intrusive laissez-faire type of regulation, whereas a Rhineland approach tends to emphasise preventive and precautionary modes of public intervention in the market. The former is associated with majoritarian political systems with high access-thresholds for new interests and issues, whereas the latter is often linked with proportional representation and a consensual tradition complemented by regular interest group consultations (Lundqvist 1980, Lehmbruch 2006, Hockerts and Schulz 2016). The respective approaches are underpinned by different legal doctrines. Roman Law featuring detailed rule-specification is often associated with the Rhineland approach, whereas Common Law with its more active judiciary is linked with the Manchester approach. Smaller European countries tend to display traits of both approaches, requiring careful analysis of their specific evolution and path-dependency.

A conventional distinction in political science between pluralism and neo-corporatism is about political processes per se, but bears some resemblance to Albert's. It is relevant because studies in the comparative environmental policy analysis literature have found that environmental performance depends on the character of political processes, contrasting pluralist to neo-corporatist democracies (Crepaz 1995, Scruggs 1999). Critics contend that the close relationship between government and peak level interest organisations of industry and labour under neo-corporatism effectively amounts to an economic growth coalition that will be mostly detrimental to environmental protection and climate mitigation. Following Scruggs (1999), the main reason why this view is probably too pessimistic is that neo-corporatist arrangements, which routinely negotiate the division of public goods, are on balance better suited to overcoming collective-action difficulties, including those relating to the environment and climate. The power of national level peak-interest organisations facilitates the pursuit of national rather than particularistic interests. Scruggs argues that once environmental problems have been duly acknowledged and accepted as relevant, neo-corporatist policymaking arrangements are adept at forging compromises effectively, committing the relevant actors to accept and provide support for the goals and means agreed. They are also smarter at finding ways to compensate potential losers in a credible way.

Pluralism implies that regulations are prepared without lawmakers' structured consultations with target groups. Consequently, pluralist systems have difficulty overcoming collective action and coordination problems to achieve good environmental outcomes, as reflected in a propensity for litigation against regulations in the courts. Vogel (1993) maintains that all western democratic political systems should be equally capable of responding to environmental challenges, if changes in the preferences of a large number of voters are intense, but concedes that in the absence of widespread public support particular characteristics of political systems can play a role. While electoral systems with proportional representation allow dedicated green interests access that is not readily available in majoritarian systems, the latter possess some other opportunity structures (Vogel 1993). There are opportunities to promote environmental policies when the executive and legislative branches are controlled by different political parties, an opportunity structure of divided government allowing for policy escalation (Lundqvist 1980). Still, majoritarian first-past-the-post electoral systems reward appeals to median voters, for whom the environment is not usually a highly salient issue (Harrison and Sundstrom 2007, p. 9).

While Scruggs (1999) and Crepaz (1995) substantiate the significance of neo-corporatism for environmental policy performance, their analyses also suggest wider influences relating to the institutional aspects of the political, legal and economic systems, including national policy styles. The decline of neo-corporatism in recent decades has directed attention to these wider framework aspects of both formal and informal institutions. Van Waarden (1995) regards neo-corporatist and consensual mechanisms as important aspects of a broader range of institutionalised practices distinct to individual countries. He defines policy style (ibid., p. 335) as 'the routine choice behavior or "standard operating procedures" which policymakers tend to develop in the policy process', as they try to cope with and reduce complexity. Routines result from lack of time and information, incomplete understanding of causal relationships, ambiguous preferences, and other constraints that limit rational decision-making. Given these uncertainties, policymakers seek to learn from previous experiences and identify appropriate rules of thumb, according to a logic-of-appropriateness.

On closer inspection, national policy styles are embedded in institutions of both state and civil society as the outcomes of long-term historical processes, often dating back to the formative phases of statehood. Given this deep-rootedness, these distinct institutions and their associated regulatory approaches, are unique to each country and not easily changed. These features need to be considered as part of a fine-grained analysis of individual countries that encompasses the design of regulations and inherent preferences for specific approaches. Consensualist traits, for instance, do not always imply neo-corporatism, as illustrated by UK preferences for a flexible and pragmatic regulatory style based on cooperation, which contrasts with the adversarial style prevalent in the USA (Vogel 1993).

The concept of national policy style was coined by Richardson *et al.* (1982, p. 13), who regarded it as resulting from the interactions between the government's approach to problem-solving and the relationship between government and other actors in the policy process. *Government's approach to problem-solving* largely covers what van Waarden refers to as routine choice behaviour, i.e. problem solutions that have proven to draw acceptable responses in the past and which tend to get repeated. For this dimension, Richardson *et al.* (1982) characterise approaches to problem-solving along a continuum from anticipatory or proactive policies to purely reactive modes. In contrast, the *relationship between government and other actors in the policy process* refers to how the government deals with interest groups in society, where a continuum runs from an accommodating policy style aimed at reaching consensus with interest groups, to decisions tending to be made and imposed notwithstanding opposition from such groups. These two dimensions effectively create a two-by-two matrix, whereby four different varieties of national policy style emerge. Some are less preoccupied with reaching consensus, as the state is seen to have a duty to enforce policies even against the wishes of organised interests, although approaches may be more or less proactive. With other policy styles, reaching consensus with interest groups is more of an imperative, even if combined with anticipatory and proactive policy traditions. Finally, some national policy styles display routine choice behaviours in a more reactive mode nevertheless combined with consensual imperatives.

Although the authors make no explicit reference to it, Richardson *et al.*'s (1982) definition, with its emphasis on the relationship between government and other actors in the policy process, provides a link to research on neo-corporatist versus pluralist approaches to policy-making and implementation. The second aspect largely covers this classical dichotomy as discussed above, whereas the first aspect emphasises the national routine intervention *modus* of government. The embeddedness of national policy styles in both state and civil society goes deeper than any individual government and somewhat beyond what can be captured under the active-reactive dimension. Van Waarden's contributions to the concept suggest that inquiries into a nation's policy style need to examine their evolution over long spans of a nation's political history, as policy styles are firmly embedded in specific and interrelated legal, political and administrative institutions that tend to reinforce and strengthen each other. Over time, a particular political culture is formed and there is a 'mutual sustainment' to a nation's political institutions, as 'culture is precipitated and embedded in legal and administrative institutions and the latter in turn buttress these cultural values, making them so enduring' (ibid., p. 361). Though incremental changes come about, any government, notwithstanding its specific political orientation, will be constrained within a path-dependent national policy style as it has evolved over time (Richardson 2017). Until a policy style becomes manifestly dysfunctional, policymakers will tend to stick to established modes of decision-making and implementation.

European integration has delivered a shared regulatory framework, but Member States have not simply converged towards one European model of policy style, as 'national decision-making procedures, coordination mechanisms and political traditions have remained essentially the same in spite of EU membership' (Liefferink and Jordan 2005, p. 111). Basically, governments continue to follow pre-existing national patterns of coordination among national actors, suggesting that national policy styles are 'sticky'.

The proposition that follows from these theoretical considerations in relation to the small countries with carbon taxation is that their national policy styles could have been conducive to their introduction. More specifically, we would expect that the small countries in question have a relationship between government and interest organisations inclined towards corporatist patterns of interest mediation, while their overall regulatory problem-solving approach is to be proactive and anticipatory, favouring coordinated market interventions.

In the next section, the characteristics of seven small EU countries that have introduced carbon taxation are explored, considering both the substantive policy processes and their political institutions, as they can be observed and classified through the lens of the policy style concept.[3] Following Scruggs (1999, p. 30) the analysis goes beyond single-country studies to enable comparison and, given the diverse set of countries, the research design is one of most-different systems design. The focus is on taxes that explicitly relate to carbon emissions. Although taxes relating to energy and transport implicitly affect emissions, they are suboptimally designed to discriminate efficiently according to emissions intensity.[4] The focus on enactment of carbon taxation implies that policy output rather than policy outcome is assumed to provide a marker of successful environmental policies.

Nordic first mover countries

Nordic governments participated in the Toronto conference and were proactive in introducing carbon taxes: in Sweden and Norway social democratic governments mandated expert commissions to review and recommend; Denmark's social-liberal Minister of Energy announced a carbon tax in his strategic policy document 'Energi 2000'; and Finland's conservative-led government took immediate action and enacted one by 1 January 1990.

Although the initial Finnish rate at €1.3/tCO$_2$ was modest, there was no mistaking the political signal.

Following Molina and Rhodes (2002) distinction between structural and actor-oriented types of corporatism, the Nordic countries practice the latter type, with strategic interdependent behaviour of trade unions, employer associations and party governments, which may or may not lead to consensus. They have a shared political history and continue their close cooperation through the Nordic Council of Ministers. According to Woldendorp's (2011) review of neo-corporatism in small countries, it has largely been continued in a business-as-usual fashion in Denmark, Norway and Finland in recent decades. Sweden has experienced some decline in corporatist practices, as conflicts in the 1980s over the creation of labour funds was followed by a breakdown of coordinated macroeconomic policies, which led to wage drift.

Sweden

The deepest economic crisis since the 1930s broke out in the early 1990s in Sweden, coinciding with the Green Party entering parliament for the first time (Lundqvist 2000). Sweden's carbon tax was enacted as part of a more encompassing fiscal reform. The unprecedented rate of the carbon tax at €80/tCO$_2$ corresponded with the European Commission's 1990 proposal. The reform simply altered the tax base for energy taxation, relating it to both carbon emissions and energy content (GJ), as Sweden had been taxing business energy use since the first oil crisis. The additional revenues from energy fuels becoming liable to VAT, as required under EU rules, enabled the tax shift that saw the lowering of payroll taxes (Andersen 1996). The carbon tax was nevertheless contested by industry and a centre-right government that came into office in 1993 lowered the rate to about €20/tCO$_2$ while altogether abolishing energy taxes for businesses. Tensions between the social partners cooled down only towards the end of the 1990s, allowing for improved coordination, but not to previous levels. As the Social Democrats returned to power, a new ambitious 10-year tax-shifting programme was launched. This time it increased carbon and energy taxes in real terms but allowed energy-intensive industries special treatment and exemptions. Further increases were suspended during a period of conservative rule, but in recent years carbon taxation has steadily increased to reach about €120 for households and services in 2018, with business now paying the same rate (EC 2018).

Finland

Finland faced economic difficulties in the early 1990s due to a banking crisis and the demise of Soviet markets causing high levels of unemployment. The carbon tax was soon complemented by an energy tax and the level of taxation increased manifold. According to one insider, the rapid increase of tax rates in the mid-1990s was largely *ad hoc* and without long-term planning (Teir 1996), although government committees had been recommending ramping up energy-related taxes for some time (Komitteebetänkande 1989). Sairinen's (2012) review of the various phases of policymaking provides detailed insights on the compromises reached in Finland, where concerns about impacts on mobility, rural areas and low-income people caused resistance to unilateral action and resulted in some exemptions for energy-intensive industries and domestic fuels. Finland's Green League played a key role, initially by putting the carbon tax on the political agenda in 1989, then by defending it and insisting on increases when the party was a member of the 1995–1999 Lipponen government (Sairinen 2012, p. 429). By 2018 the tax had reached €73/tCO$_2$.

Denmark

The political process leading to the adoption of Denmark's carbon tax (by 2018 spanning €10–23/tCO$_2$) was exceptional in that a majority in parliament insisted on the tax against the will of the ruling (minority) coalition government. As the Social Democrats came into power in 1993, the further development of carbon-energy taxation as part of environmental tax reforms, catalysed by a smaller coalition partner,[5] was in line with business-as-usual processes of interest group concertation, as three consecutive fiscal bills were agreed from 1995 to 1999. Business had up to then not been subject to energy taxes. The Danish tax system is largely devoid of payroll taxes and social contributions from employers, so other mechanisms for compensation had to be developed (Klok *et al.* 2006). Unlike Sweden and Finland, Denmark decided to recycle 20 per cent of the revenues for energy efficiency measures, which supported adjustment processes in all parts of business life (Andersen and Speck 2009). Following the shift to a liberal-conservative coalition in 2001, tax increases were put on hold for several years, but in 2009 this coalition reduced income taxes against adjustments and indexation of carbon and energy taxes.

Overall policy styles in the Nordic countries reflect a proactive problem-solving approach and a consensual tradition. Parliamentary institutions feature proportional representation, with an extended accommodation of political minorities through mechanisms that distribute additional seats to mirror closely the share of votes achieved at the national level for each political party. It allows ecologically oriented parties to win seats and gain influence by directly or indirectly participating in government, while having precluded single-party majoritarian governments during this period. The political processes leading to adoption of carbon taxes have been conflictual and with more stop-and-go processes involved than one would perhaps expect from the political science literature that tends to downplay the bargaining and compromising involved with consensual approaches (e.g. Molina and Rhodes 2002). The frequency of special arrangements for energy-intensive industries reflects the ability to strike deals and find means to compensate potential losers, although it has also watered down potential climate mitigation effects (Ekins and Speck 2000). Nevertheless, the patterns of policymaking on carbon taxation are broadly in line with theoretical expectations for relationships between neo-corporatist modes of policymaking and implications for policy outputs.

Countries in transition: Slovenia and Estonia

Surprisingly, it was a small country with an economy in transition that was the first outside the Nordic region to adopt a carbon tax. In 1997 Slovenia exchanged an ad-valorem energy tax with nominal fuel tax rates and a broadening of the tax base involving a carbon tax differentiated according to the specific carbon emissions of fossil fuels. Two more transition economies subsequently followed suit: Estonia in 2000 and Croatia in 2007.

Countries in transition represent a moving target regarding national policy styles, as novel political institutions based on democracy and accountability need a longer time span to define themselves and their *modus operandi*. With a relatively blank slate, learning processes are required for policymakers to identify successful routines for problem-solving. Considering that the countries prior to World War II were briefly independent, democratic and free market-oriented entities, significant path-dependencies are in play too.

Slovenia

Although Slovenia is a young democracy, political scientists agree that 'corporatism has been an influential doctrine in the Slovenian polity since its beginning', as political life has shown a strong inclination to organise around interest groups from the time of the disintegration of the Austro-Hungarian Empire (Lukšič 2003, Haček 2009). These policy style traits persisted during the post-war Yugoslavian republic, where Slovenia was the most economically advanced country of the federation and Slovenian intellectual Edvard Kardelj played a key role in developing Yugoslavia's distinctive self-management democracy. In the present-day Slovenian polity, a structured form of corporatism is embodied in the 'National Council', the second parliamentary chamber consisting of representatives from various interest groups.

A steep increase in emissions following a normalisation of the economy triggered Slovenia's carbon tax in a context of fiscal needs (Markovič-Hribernik and Murks 2006). Extensive negotiations conducted by a designated expert group preceded its adoption. Once a formula had been agreed, the subsequent parliamentary approval was relatively smooth, according to Peter Novak, former chairman of Slovenia's CO_2 tax committee (personal communication 19 May 2016). The approach adopted included certain reductions for energy-intensive industries and the hypothecation of revenues for climate mitigation measures (Markovič-Hribernik and Schlegelmilch 1999). Fiscal challenges with high inflation rates during the years from adoption to Slovenia's accession to the EU paved the way for subsequent tax rate increases, whereby today's rate of €18/tCO_2 was reached, apparently without further controversy over its role in the tax system. Slovenia's Green Party was partner in the broad-based coalition governments from 1990–1996, but seems to have had no direct role in devising the carbon tax.

Estonia

Like Slovenia, Estonia was an economically advanced entity within a wider union and proactively pioneered a number of the free market reforms initiated under Gorbachev in the Soviet Union. With Finnish as its second language, it benefits from its proximity to the Nordic countries. Estonia's persistent quest for independence was reflected in the government-in-exile maintained during the entire Soviet era. Unlike Slovenia, no successor party to the Communists emerged, and the government has continuously been firmly center-right with a strong emphasis on economic liberalisation. While Estonia is oriented towards an Anglo-Saxon style of problem-solving, the legacy from the earlier period of independence (1918–1940) has provided the country with a political system based on proportional representation (Panagiotou 2001, Adam *et al.* 2009).

Taxation policy was from the outset based on the principle of a low, flat tax rate for all personal and corporate income, and the carbon tax came into play with efforts to lower the flat tax to 20 per cent (Pettai 2009, p. 77). It was only in 2005–6 that a carbon tax was promoted along with other green taxes under the banner of environmental tax reform, at a time when Estonia's Green Party had surged in the polls (Sikk and Andersen 2009). Starting from a low level, the tax rate has gradually increased, reaching €2/tCO_2, and with intentions to align it to the emissions trading system (ETS) allowance price. With a relatively limited share of traditional Soviet-style heavy industries in Estonia, the competitiveness impact of the tax has been limited (Kiuila and Markandya 2009). Reflecting routines of problem-solving, coalition-building and compromise in the parliamentary realm, rather than with interest organisations, seems to have enabled the adoption of the tax.

According to Adam *et al.* (2009) the policy styles and transition paths of Slovenia and Estonia have diverged, the latter favouring a radical approach to problem-solving and the former an incremental approach. Nevertheless, the political institutions of both countries feature proportional representation, accommodating consensus-seeking and a catalysing role by a Green party. Slovenia experienced a more consensual, incremental policy that added carbon taxation into preexisting energy tax structures, and involving negotiations and compromises with relevant interest organisations. Estonia departed in a more radical way by imposing novel tax policies featuring carbon taxation. For both countries carbon taxes anticipated upcoming EU membership and demonstrated commitment to European policies regarding climate and financial stability, even if there was leeway from high emission baselines of the past. The routines of problem-solving can thus be characterised as proactive in both Slovenia and Estonia, while policy styles diverge with regard to the government relationship to other actors in the policy process.

Cohesion countries: Ireland and Portugal[6]

In 2010, Ireland became the first of the initial cohesion countries to introduce a carbon tax. Despite 2010 being an *annus horribilis* with the cost of bailing out Irish banks rising to one-third of GDP, the carbon tax had been agreed in government coalition bargaining in advance of the 2008 fiscal and debt crisis (Convery *et al.* 2013). In Spain, a government-appointed tax reform commission presented a carbon tax proposal in 2013, but it was Portugal that in 2014 became the second cohesion country to unilaterally adopt a carbon tax, following recommendations from an expert commission (Commissão para a reforma da fiscalidade verde 2014).

The Cohesion Fund was established to strengthen the economic, social and territorial cohesion of the EU in the interests of promoting sustainable development, and is reserved for Member States whose gross national income (GNI) per capita is less than 90 per cent of the EU average. Three of the four first-generation cohesion countries owe their economic under-development partly to their totalitarian past, as Portugal, Spain and Greece were fascist dictatorships with command economies well into the mid-1970s. Despite sharing some clientelist traits with these countries, Ireland is often characterised as a pluralist country based on the Anglo-Saxon model (Hall and Soskice 2001, Ó'Riain 2014).

Ireland

Woldendorp (2011) shows how Ireland prior to the fiscal crisis and over a period of 20 years had seen an uninterrupted chain of national agreements on wages and taxes between government, trade unions and employers, reflecting a social partnership very much in line with conventional neo-corporatist policy styles. There is no encompassing welfare state or strong labour party, but well-established patterns of corporatist political exchange had emerged – partly in continuation of a policy style of 'brokerage' dating back to the early years of independence (Ó'Riain 2014, p. 173) – only to be suspended by 2009 (Moran 2010). Ireland has, unlike the majoritarian model prevalent in Anglo-Saxon countries, an electoral system of proportional representation with a single transferable vote, which has allowed a small Green party to emerge and in 2007 to become a coalition partner in Ireland's government.

The Greens had campaigned on introducing a carbon levy, and the coalition government introduced a tax at the rate of €15/tCO$_2$. Emission projections indicated that without it Ireland would overshoot on its climate policy obligations by 3–5 million tonnes by 2020 (ibid., p. 341). Negotiations with the Troika (European Commission, IMF and European Central Bank) caused exemptions, e.g. for coal, to be abolished and a rate increase to €20/tCO$_2$ for the purposes of fiscal consolidation. Vehicle taxes were also aligned to CO$_2$-emissions, strengthening the overall emphasis on taxation policies for climate mitigation.

Portugal

Portugal became democratic in 1974 following a peaceful revolution that resulted in a polity with Socialists and (centre-right) Social Democrats in the dominating roles. The electoral system is based on proportional representation, allowing representation of several parties – including Ecologists – in the National Assembly. Formal corporatist institutions existed during the dictatorship (1926–1974), featuring a distinct chamber of the National Assembly with interest organisations. They were abandoned with the advent of democracy and the freedom to form independent labour unions (Gallagher 1983, p. 62). World market oil price increases around 2000 contributed to the slowing of Portugal's economy well before the present crisis when economic growth declined below the EU average. Portugal embarked upon an energy transition, resulting in renewables now accounting for over half the electricity supply. Nevertheless, GHG emissions have increased by 13 per cent from 1990.

Portugal has a large external debt that was subjected to much higher interest rates following the financial crisis, creating a dangerous fiscal squeeze on the economy in 2011. This situation opened a window of opportunity for the introduction of a carbon tax, but it emerged only after a range of other painful measures had been exhausted. The rescue plan agreement with the Troika included a cut in retirement pensions, which Portugal's High Court ruled unconstitutional. This decision left a gap of several billion Euros to consolidation requirements. A blueprint of a potential environmental tax reform (EEA 2013) gave support to an activist minister and encouraged the centre-right government to establish a Green Tax Reform Committee of experts, which proposed a range of new taxes, including one on carbon. The tax rate ($€6.9/tCO_2$ by 2018) was designed to reflect the ETS allowance price. The green tax package was largely revenue-neutral, allowing for deductions in income taxes according to the number of dependent children and seniors (Governo de Portugal 2014). The intention may have been to sweeten the pill of harsh austerity measures, but it could not prevent the centre-right government losing the 2015 election.

According to Manuel (2010, p. 8), the fathers of Portugal's revolution railed against eastern European 'socialist misery' preferring to look to problem-solving approaches in Sweden and Germany for inspiration. Whereas Portugal undertook a decisive break with its previous dysfunctional institutions and policy styles, Ireland made a softer departure from British-style majoritarian approaches. The overall tradition of problem-solving in both countries tends towards an incremental policy style, limited by legacies of poor economic performance and troubled pasts. Ireland's carbon tax reflected a short-lived run-up towards a more proactive taxation policy during the 'Celtic tiger' years (Commission of Taxation 2009, Ó'Riain 2014), in contrast to Portugal's tax that came about as a result of fairly reactive processes. Both governments acted in the shadow of proportional representation dynamics, while the international financial crisis and the Troika further catalysed policy-making processes by creating willing Departments of Finance within the government.

Discussion

This concise review of the adoption of carbon taxes in seven small countries shows how circumstances and policy frameworks have differed. Still, such taxes frequently become part of wider fiscal packages that help blur the clear distinction between winners and losers that would follow from a stand-alone carbon tax.

At the time of adoption, several of the small countries had energy systems with relatively high shares of carbon-neutral energy carriers, which lowers political resistance. Sweden, Slovenia and Portugal benefit from significant hydropower resources. Other countries were locked into fossil fuels, so a carbon tax could be expected to create higher economic as well as political costs. Denmark, Finland, Ireland and Estonia relied significantly on coal and other fossil fuels. Closer inspection of tax bands and exemptions to energy-intensive industries reveal implications of the share of low carbon fuels for the level of ambition in carbon taxation (see Andersen 2015), but there is no clear-cut relationship as reflected in the relatively substantial tax rates in Ireland and Finland. Small countries have adopted carbon taxes despite the nature of their energy systems and have gone beyond symbolic measures.

Carbon taxation is advocated by green parties, but in our sample of small countries their role was not always explicit. It is mainly in Ireland and Finland that green parties were able to achieve the tax or rate increases as a policy concession, whereas green or green-ish parties seem to have provided a catalysing effect to governments in other small countries. The entrepreneurs in the core group of Scandinavian countries have been social democratic coalition governments, which introduced carbon taxes as part of wider macroeconomic policy packages in wheelings and dealings with business and labour interests. In Portugal, Ireland and Estonia, and to some extent in Finland, the carbon tax was, in contrast, championed by centre-right coalition governments.

These observations suggest that the feasibility of carbon taxation is not influenced by the nature of energy systems or party political leadership in any simple or straightforward way. Following the theoretical debate reviewed above, the broader features and varieties of the national policy styles of small countries attract attention, i.e. the institutionalised approaches to problem-solving and the relationship between government and other actors in the policy process. The comparative environmental policy literature makes a strong case for the significance of neo-corporatism to the effectiveness of environmental policies. It views neo-corporatist policy arrangements as a necessary protective device for small countries with open and vulnerable economies. Despite the decline of neo-corporatism since the 1970s, policymaking routines involving negotiations with interest groups remained in place and provided a suitable vehicle for policies involving carbon taxation. Traits of interest concertation were also observed in small countries outside the traditional core group of Nordic countries. Legacies from structurally-oriented varieties of corporatism have provided support for consensual mechanisms of policymaking in Slovenia and Portugal, as their national policy styles undergo incremental transformations. Developments in Ireland towards social partnership from 1987–2007 show that such patterns could unfold in small countries with a policy style shaped by a different past. Indeed, evolving practices in Portugal and Ireland were identified early on by Katzenstein (2003, p. 23). While Slovenia, Ireland and Estonia all display traits of proactive policy styles in carbon taxation, Portugal acted according to slightly more reactive routines of problem-solving.

The adversarial nature of policymaking in pluralist political systems is reflected not only in the challenging conditions facing carbon taxes in the USA, but also in the fate of the recent Australian Carbon Pricing Mechanism. Promoted by the Greens, who made use of their pivotal position in the Australian Senate to make adoption of a carbon pricing scheme a condition of their support for a minority Labor government, it was repealed following the return of a conservative majority government (Rootes 2014, Ward 2015). The introduction of carbon taxes in small European countries was controversial in many instances too, however, their policy processes over time involve considerable bargaining and compromise that gradually commits all the main actors to the deals made. Carbon taxes, along with energy taxes, generate significant revenues and make up for other taxes that might be equally unpopular to increase. Under these circumstances, the short-term political costs of continuing carbon taxes are small, although rates or exemptions can be renegotiated.

Thus, the reason why small countries have a greater propensity to introduce carbon taxes seems to be their varieties of national policy styles in which fiscal and macroeconomic policies become subject to intense coordination and negotiation, providing better opportunities for introducing such taxes as part of wider redistribution mechanisms. Indeed, the concept of environmental fiscal reform, whereby labour taxes are cut in exchange for environmentally-related taxes, offers the opportunity to lower labour costs by reducing employers' social contributions. As these are in most cases unitary states, the number of institutional veto points is limited once the government can establish a majority compromise. Electoral systems with proportional representation offer, as Harrison (2010) notes, better access for supporters of unilateral climate mitigation policies, notably parties of green orientation, which can act as facilitators from their position in the political system. In addition, small countries are by virtue of their size often more homogenous overall, limiting the complexity of credible compensation arrangements.

Neo-corporatist systems with formal access to policymaking for regional interests have had more mixed histories. Two countries without carbon taxation often classified as neo-corporatist and core to Rhineland capitalism, Austria and Germany, are federal polities in which there is formal access for regional interests to influence policymaking, via a Federal Council. Even if there is no divided government, individual states (Länder) can join the opposition on specific issues, delaying or blocking approval. Federal countries tend to move forward more slowly, in line with Scharpf's (1988) concept of the joint decision trap. In response to the question of why Germany has no explicit carbon tax, despite its ecological tax reform and high profile on climate policy, former Finance Minister Hans Eichel (SPD) of the red-green coalition replied: 'The political costs of introducing a completely new tax are too high' (personal communication 12 July 2012). Germany, like the Netherlands, instead opted for increases in excise taxes, whereby coal is favoured. A renewed attempt to introduce a climate tax in Germany failed due to opposition from a 'broad coalition of trade unions, industry, brown coal states, Bavarian CSU and eastern parts of SPD', backed by legalistic arguments (Spiegel 2015, Welt am Sonntag 2015).

The national policy styles in Germany and Austria incline towards rule-making, shaped by the path-dependency of ordo-liberalism (Bonefeld 2012). Equal treatment under the rule of law is in jeopardy from carbon taxes that by definition allow for differentiated responses. Yet, Switzerland, a small, multi-linguistic state featuring a comparable regulatory culture, has managed to enact and ramp up a carbon tax by making its implementation conditional on whether the rules – greenhouse gas reduction targets – were complied with (Gerigk et al. 2012).

Conclusions

Here, I have extended previous research on the role of neo-corporatism for environmental policy performance into an inquiry on the significance of varieties in national policy styles. The concept of national policy style addresses a government's routine choice behaviours with its relationship to the main actors in the policymaking process. Routine choice behaviours are formed over time as a result of learning processes and the need of policymakers to deal with limited time and resources. This creates coping strategies and political cultures that become embedded in political and legal institutions. Even countries with dysfunctional policy styles that seek to change course are subject to inertia and enduring approaches.

Previous studies of climate policy instrument choices have considered contemporary policy drivers to further multitrack designs for international agreements. Analyses of emission developments and climate policy emulations have nevertheless led to observations that there are underlying 'important background features within the domestic structure of countries that the international climate change negotiations are unlikely to change, at least in the short term' (Lachapelle and Paterson 2013, p. 565).

Here, I have restricted myself to analysing the adoption of carbon taxes in order to improve our understanding of some of the mechanisms at play. I find that the small countries have often adopted carbon taxation due to their basic routines for proactive problem-solving policies, with decision-making that involves consultations between government and the main economic interests an integral part of prevailing national policy styles.

These patterns of interest mediation enable coordination of fiscal and macroeconomic policies with reasonably credible mechanisms for compensating potential losers, thus overcoming political resistance. Exceptions could be observed in Portugal, reflecting a national policy style of more reactive problem-solving although combined with consensus-seeking, and Estonia, reflecting a national policy style of more top-down impositional decision-making although combined with routines for proactive problem-solving approaches. However, the other five countries display broadly comparable traits. In all seven countries carbon taxation has entered policy-making through fiscal reform rather than through climate mitigation policy *per se*, which confirms previous suggestions that carbon taxation's feasibility in fiscal policy vis-à-vis other taxes is better than its feasibility in climate policy vis-à-vis other policy instruments (cf. Jagers and Hammar 2009).

Despite these observations, the role of neo-corporatist routines has overall been somewhat weaker than expected, with consultations and negotiations between government and stakeholders not always involving peak-level organisations. Neo-corporatist modes of decision-making have been in decline in several countries with the most distinctive traits, such as Sweden. While research suggests the presence of neo-corporatist patterns in other small countries, it is by no means as structured or formalised as might have been expected. The legacies of neo-corporatism as well as of old style corporatism (as in Portugal and Slovenia) rather seem to provide a cultural norm for policy-making in which it is routine for government to consult in a structured way with economic interests and in which government is expected to compromise and broker its policies by offering some compensation.

The analysis reveals a stronger influence of the character of political institutions than expected and confirms observations elsewhere relating to the significance of proportional representation and coalition governments (Harrison 2010, Lachapelle and Paterson, 2013). The ability of parties supporting unilateral climate mitigation policies to make it into government helps sensitise larger parties. These observations are in line with Scruggs (1999, p. 6) who states that 'parties under proportional representation may gain policy influence through the process of party competition and compel larger parties to accommodate those interests'. With traditional parties accustomed to neo-corporatist patterns of interest mediation, the dynamics of voter preferences help further political exchanges structured by institutions and problem-solving approaches of preexisting national policy styles. A 'mutual sustainment' between political institutions and the political culture associated with national policy styles indeed seems to be at work, as suggested by van Waarden.

An explanation of why small countries have been more capable of taxing carbon unilaterally, despite being more open to forces of international competition, thus draws on a complex tale of how their political institutions shape national policy styles, understood as a government's relations to the main economic actors and its routines for problem-solving as they have evolved over time.[7]

Notes

1. A carbon tax is defined as a tax levied on the carbon content of fuels (Yokoyama *et al.* 2000).
2. A carbon price floor for emission allowances introduced in the UK (2013) applies to electricity only and is not an economy-wide carbon tax scheme as such. Poland's levy on carbon is part of a general air pollution tax and not a distinct carbon tax, and Ukraine's CO2 charge is set at a symbolic rate of about USD 0.02/tCO2.
3. Norway, Iceland and Switzerland are not EU Member States, while Croatia only recently became so (2013).
4. For a detailed review of tax bases and tax rates see Andersen (2015) and OECD (2016).
5. The Socialist People's Party, which 'has been in many respects a traditional green party, with a participatory and egalitarian culture and a strong emphasis on environmental policies' (Kosiara-Pedersen and Little 2016).
6. The Cohesion Fund, set up in 1994 by Council Regulation (EC) 1164/94, provides funding for environmental and trans-European network projects. The first generation of countries eligible for support included Greece, Ireland, Portugal and Spain.
7. Being more likely to enact carbon taxes is, however, a relative phenomenon. With the Paris Agreement the world has moved beyond unilateral action, prompting larger countries, such as Mexico, Chile and South Africa, to announce carbon taxes.

Acknowledgments

Participants at the 'Climate Politics in Small States' workshop at Dublin City University June 2016 and the INOGOV workshop 'Pioneers and Leaders in Polycentric Climate Governance' Hull (UK) September 2016 provided stimulating comments and suggestions. I am grateful to two anonymous reviewers and the special issue editors for constructive and insightful remarks.

Disclosure statement

No potential conflict of interest was reported by the author.

Funding

This research did not receive any specific grant from funding agencies in the public, commercial or not-for-profit sectors.

References

Adam, F., Primoz, K., and Tomsič, M., 2009. Varieties of capitalism in Eastern Europe (with special emphasis on Estonia and Slovenia). *Communist and Post-Communist Studies*, 42, 65–81. doi:10.1016/j.postcomstud.2009.02.005

Albert, M., 1993. *Capitalism against capitalism*. London: Whurr publishers.

Albert, M. and Gonenc, R., 1996. The future of Rhenish capitalism. *Political Quarterly*, 67 (3), 184–193. doi:10.1111/poqu.1996.67.issue-3

Andersen, M.S., 1996. The domestic politics of carbon-energy taxation. *In*: M. S. Andersen and D. Liefferink, eds.. *The new member states and the impact on environmental policy. Final report to EC-DGXII*. Aarhus: Aarhus University, 111–140.

Andersen, M.S., 2015. Reflections on the Scandinavian model: some insights into energy-related taxes in Denmark and Sweden. *European Taxation*, 55 (6), 235–244.

Andersen, M.S. and Speck, S., 2009. Energy-intensive industries: mitigation and compensation. *In*: M.S. Andersen and P. Ekins, eds.. *Carbon-energy taxation*. Oxford: Oxford University Press, 120–143

Bonefeld, W., 2012. Freedom and the strong state: on German ordoliberalism. *New Political Economy*, 17 (5), 633–656. doi:10.1080/13563467.2012.656082

Commissão para a reforma da fiscalidade verde, 2014. *Projeto de reforma da fiscalidade verde*. Lisboa, Governo de Portugal, p. 293.

Commission on Taxation (An Coimisiún um Chánachas), 2009. *Report*. Dublin: The Stationery Office.

Toronto Conference, 1988. The changing atmosphere: implications for global security. *In*: D.E. Abrahamson, ed. *The challenge of global warming*. Wash. DC: Island Press, 44–62.

Convery, F.J., Dunne, L., and Joyce, D., 2013. Ireland's carbon tax and the fiscal crisis. *OECD working papers*, no. 59. Paris.

Crepaz, M., 1995. Explaining national variations of air pollution levels: political institutions and their impact on environmental policy-making. *Environmental Politics*, 4 (3), 391–414. doi:10.1080/09644019508414213

EEA, 2013. *Environmental fiscal reform – illustrative potential in Portugal*. Copenhagen: European Environment Agency, Staff Position Note 13/01

Ekins, P. and Speck, S., 2000. Proposals of environmental fiscal reforms and the obstacles to their implementation. *Journal of Environmental Policy and Planning*, 2 (2), 93–114. doi:10.1080/714038548

Gallagher, T., 1983. *Portugal: A twentieth-century interpretation*. Manchester: Manchester University Press.

Gerigk, J., *et al.*, 2012. *The current climate and energy policy in the EU and Switzerland*. Zürich: Swiss Federal Institute of Technology.

Governo de Portugal, 2014. *The green growth commitment and the green taxation reform*. Lisbon: Ministry of Environment, Spatial Planning and Energy.

Haček, M., 2009. Understanding politics in Slovenia. *In*: L. Johannsen and K. Hilmer Pedersen, eds.. *Pathways*. Aarhus: Aarhus University Press, 98–116

Hall, P. and Soskice, D., eds. aug 2001. *Varieties of capitalism: the institutional foundations of comparative advantage*. Oxford: Oxford University Press

Harrison, K., 2010. The comparative politics of carbon taxation. *Annual Review of Law and Social Science*, 6, 507–529. doi:10.1146/annurev.lawsocsci.093008.131545

Harrison, K. and Sundstrom, L.M., 2007. The comparative politics of climate change. *Global Environmental Politics*, 7 (4), 1–18. doi:10.1162/glep.2007.7.4.1

Hockerts, H.G. and Schulz, G., eds, 2016. *Der 'Rheinische Kapitalismus' in der Ära Adenauer*. Paderborn: Ferdinand Schöningh.

Jagers, S.C. and Hammar, H., 2009. Environmental taxation for good and for bad: the efficiency and legitimacy of Sweden's carbon tax. *Environmental Politics*, 18 (2), 218–237. doi:10.1080/09644010802682601

Katzenstein, P.J., 2003. Small states and small states revisited. *New Political Economy*, 8 (1), 9–30. doi:10.1080/1356346032000078705

Kiuila, O. and Markandya, A., 2009. Can transition economies implement a carbon tax and hope for a double dividend? The case of Estonia. *Applied Economics Letters*, 16, 705–709. doi:10.1080/13504850701221816

Klok, J., *et al.*, 2006. Ecological tax reform in Denmark. *Energy Policy*, 34, 905–916. doi:10.1016/j.enpol.2004.08.044

Komitteebetänkande, 1989. *Ekonomisk styrning af miljövärden*. Helsinki: Statens Tryckericentral.

Kosiara-Pedersen, K. and Little, C., 2016. Environmental politics in the 2015 Danish general election. *Environmental Politics*, 25 (3), 558–563. doi:10.1080/09644016.2015.1123825

Lachapelle, E. and Paterson, M., 2013. Drivers of national climate policy. *Climate Policy*, 13 (5), 547–571. doi:10.1080/14693062.2013.811333

Lehmbruch, G., 2006. Nationen und Systemtypen in der vergleichenden politischen Ökonomie. *In*: V.R. Berghahn and S. Vitols, eds. *Gibt es einen deutschen Kapitalismus?* Frankfurt a.M.: Campus, 86–95.

Liefferink, D. and Jordan, A., 2005. An 'ever closer union' of national policy? *European Environment*, 15, 102–113. doi:10.1002/eet.377

Luksič, I., 2003. Corporatism packaged in pluralist ideology: the case of Slovenia. *Communist and Post-Communist Studies*, 36, 509–525. doi:10.1016/j.postcomstud.2003.09.007

Lundqvist, L., 1980. *The hare and the tortoise: clean air policy in the United States and Sweden*. Ann Arbor: The University of Michigan Press.

Lundqvist, L., 2000. Capacity-building or social construction? Explaining Sweden's shift towards ecological modernization. *GeoForum*, 31, 21–32. doi:10.1016/S0016-7185(99)00041-X

Manuel, P.C., 2010. Portuguese exceptionalism and the return to Europe. *In*: L. C. Ferreira-Pereira, ed.. *Portugal in the European Union*. London: Routledge, 15–29.

Markovič-Hribernik, T. and Murks, A., 2006. The long road from Ljubljana to Kyoto. *Financial Theory and Practice*, 30 (1), 29–65.

Markovič-Hribernik, T. and Schlegelmilch, K., 1999. Green budget reform in Slovenia. *In*: K. Schlegelmilch, ed. *Green budget reform in Europe*. Berlin: Springer, 293–400.

Molina, O. and Rhodes, M., 2002. Corporatism: the past, present and future of a concept. *The Annual Review of Political Science*, 5, 305–331. doi:10.1146/annurev.polisci.5.112701.184858

Moran, J., 2010. From Catholic Church dominance to social partnership promise and now economic crisis. *Irish Journal of Public Policy*, 2, 1.

Ó'Riain, S.O., 2014. *The rise and fall of Ireland's Celtic Tiger*. Cambridge: Cambridge University Press.

OECD, 2002. *Environmental performance reviews: italy*. Paris: OECD.

OECD, 2016. *Effective carbon rates: pricing CO2 through taxes and emissions trading systems*. Paris: OECD.

Panagiotou, R.A., 2001. Estonia's success. *Communist and Post-Communist Studies*, 34, 261–277. doi:10.1016/S0967-067X(01)00005-8

Pearce, D., 2006. The political economy of an energy tax: the UK's climate change levy. *Energy Economics*, 28, 149–158. doi:10.1016/j.eneco.2005.10.001

Pettai, V., 2009. Understanding politics in Estonia. *In*: L. Johannsen and K. Hilmer Pedersen, eds.. *Pathways*. Aarhus: Aarhus University Press, 69–97

Richardson, J., 2017. The changing British policy style. *British Politics*, 12 (2), 1–19.

Richardson, J., Gustafsson, G., and Jordan, G., 1982. The concept of policy style. *In*: J. Richardson, ed. *Policy styles in Western Europe*. London: Allen & Unwin, 1–16.

Rootes, C., 2014. A referendum on the carbon tax? *Environmental Politics*, 23 (1), 166–173. doi:10.1080/09644016.2014.878088

Sairinen, R., 2012. Regulatory reform and development of environmental taxation. *In*: M.S. Andersen and J. Milne, eds. *Handbook of research on environmental taxation*. Cheltenham: Edward Elgar, 422–438.

Scharpf, F., 1988. The joint decision-trap. *Public Administration*, 66 (3), 239–278. doi:10.1111/j.1467-9299.1988.tb00694.x

Scruggs, L., 1999. Institutions and environmental performance in 17 western democracies. *British Journal of Political Science*, 29 (1), 1–31. doi:10.1017/S0007123499000010

Sénit, C., 2012. The politics of carbon taxation in France. *IDDRI working paper* 20, Paris: SciencePo.

Sikk, A. and Andersen, R.H., 2009. The fall and rise of Estonian Greens. *Journal of Baltic Studies*, 40 (3), 349–373. doi:10.1080/01629770903118740

Speck, S., 2009. Annex: effective carbon-energy tax rates. *In*: M.S. Andersen and P. Ekins, eds.. *Carbon-energy taxation*. Oxford: Oxford University Press, 282–302

Spiegel, 2015. Koalition beerdigt Klimaabgabe. *Spiegel*, 2 July .

Teir, G., 1996. Evolution of CO2/energy taxes in Finland. *In*: European Foundation for the Improvement of Living and Working Conditions, ed. *Environmental taxes and charges*. Luxembourg: Office for Official Publications of the European Communities, 243–248.

van Waarden, F., 1995. Persistence of national policy styles: A study of their institutional foundations. *In*: B. Unger and F. van Waarden, eds. *Convergence or diversity*. Aldershot: Avebury, 333–371.

Vogel, D., 1993. Representing diffuse interests in environmental policy-making. *In*: R.K. Weaver and B.A. Rockman, eds. *Do institutions matter ?* Washington, DC: The Brookings Institution, 237–271.

Ward, I., 2015. The campaign against the carbon tax, the media and a new uncivil politics. *Australian Journal of Political Science*, 50 (2), 225–240. doi:10.1080/10361146.2015.1005571

Welt am Sonntag, 2015. Juristen zerpflücken Gabriels Kohlepläne. *Welt am Sonntag*, 3 May .

Woldendorp, J., 2011. Corporatism in small north-west European countries 1970–2006. *Working Paper Series, Dept. of Political Science, VU University Amsterdam*, no. 30.

Yokoyama, A., Ueta, K., and Fujikawa, K., 2000. Green tax reform: converting implicit carbon taxes to a pure carbon tax. *Environmental Economics and Policy Studies*, 3 (1), 1–20. doi:10.1007/BF03353964

Zito, A., 2002. Integrating the environment into the European Union: the history of the controversial carbon tax. *In*: A. Jordan, ed. *Environmental policy in the European Union*. London: Earthscan, 241–255.

The Czech Republic's approach to the EU 2030 climate and energy framework

Mats Braun ⓘD

ABSTRACT

The East-West divide within the EU over climate policy has been frequently discussed. There is a tendency in the literature to focus on Poland and ignore the other countries in the central and eastern European region. Here it is argued that the institutionalised cooperation between the four countries in the Visegrad Group (the Czech Republic, Hungary, Poland and Slovakia) provides a crucial component for an understanding of how the participating countries approach EU climate negotiations. Here it is suggested that the group is important as a bargaining coalition but also as a reference point for the development of shared 'Visegrad' norms in the field. This is based on a case study of the Czech Republic's approach to the 2014 negotiations on the 2030 climate and energy framework and the country's cooperation with the other Visegrad countries on the issue.

Introduction

The role of the central and eastern European member states in EU climate and energy policy has received some attention in the scholarly literature, primarily focused on the role of Poland and the country's negative influence on the possibility of the EU adopting a progressive climate policy (see, e.g. Skovgaard 2013, Skjærseth 2016). The discussion acquired renewed relevance after the negotiations of the 2030 EU climate and energy policy framework in 2014. The Polish Prime Minister Ewa Kopacz then declared her country to be the winner of the EU negotiations since the deal struck would not, in her view, lead to increased energy prices in Poland (Darby 2014). Less attention, however, has been paid to the role the smaller Visegrad countries play in the development of EU climate policy. For these countries the institutionalised cooperation within the Visegrad Group (V4) was a crucial component of their approach to the negotiation of the EU 2030 framework. The V4 countries managed to coordinate their positions during those negotiations, and several of the group's initial demands were reflected in the final package agreed upon by the European Council in October 2014. In particular the group managed to find support for their vision of the compensatory mechanism favouring low income countries, the extension of the derogation from auctioning for parts of the energy generating sector, and the specific statements regarding the role of the European Council (as opposed to the Council) in reviewing the future development of the policy framework (Joint Statement 2014; Conclusions on 2030 Climate and Energy Policy Framework). Therefore, representatives of the group considered this a success of their collaboration (see, e.g. Prouza 2015).

The success of the Visegrad cooperation on the issue of EU climate and energy policy suggests that researchers should take this form of cooperation seriously despite the limited levels of formal institution building within the V4. I argue that 'institutions matter' (Tallberg 2010, p. 634) also in the case of the sub-regional groupings. In the extensive scholarly literature on the EU only very little has been written on sub-regional groupings within the EU, such as the Visegrad Group, the Nordic-Baltic 6, and the Benelux group (for some exceptions see Rüse 2014, Fawn 2013, Dangerfield 2008, Cottey 2009).

Here, I discuss the relationship between a member state, a sub-regional group and a broader unit of regional integration, i.e. the Czech Republic, the V4 and the EU. I discuss how we should understand the role of the V4 in the climate negotiations in relation to the V4 member states. I do so through a detailed analysis of the V4 activities during the negotiations and of how one of the smaller V4 countries, the Czech Republic, perceived its participation in the V4 coalition on the issue. I ask which role the V4 cooperation has played in the development of the Czech approach in the EU climate negotiations. I therefore speak to the literature on sub-regional groupings within the EU. However, there is also the empirical relevance of studying the V4 activities regarding climate policy for anyone interested in the EU's climate policy.

I argue in line with previous research that the V4 can be understood as a territorially constituted coalition (Rüse 2014), but stress that this is not the full story. For a better understanding of the internal dimension of the V4 I utilise two concepts originating in the sociological 'new institutionalist' literature: rhetorical action (Schimmelfennig 2001) and socialisation (see, e.g. Checkel 2007). The conclusion suggests that the V4 group's appeal to economic fairness corresponds to the idea of rhetorical action since the demands for a burden-sharing agreement reflect a shared EU norm of economic solidarity. At the same time, the group's approach has been developed as a response to and in conflict with ideas of ecological moder-nisation that legitimise EU climate policy. The V4 cooperation and the frequent meetings between representatives provide for a potential platform for a process of socialisation that in this case reinforces norms that contradict the Europeanisation process.

I have based this discussion on an analysis of official materials from the V4 and the Czech Republic as well as interviews with Czech civil servants from the Office of the Government, the Ministry of Environment and the Ministry of Industry and Trade.[1]

In the first part, I outline the approach based on new institutionalism. In the second part, I present an overview of the activities and institutional structure of the V4. Thereafter, I turn to analysing the V4's approach to the EU's climate and energy policy, and follow this with a more detailed examination of the Czech approach to the 2030 climate and energy policy framework. I then devote the last section to a discussion on the two key concepts – rhetorical action and socialisation – and their relevance for our understanding of the V4 and its approach to EU climate and energy policy.

Sub-regional integration in the EU and new institutionalism

There is relatively little literature on sub-regional groupings within the EU (for some exceptions see Rüse 2014, Fawn 2013, Dangerfield 2008, Cottey 2009). This literature has often focused on the role of these groups within the EU from a perspective of coalition building and rational choice institutionalism. From this perspective, the groups have been labelled 'territorially constituted coalitions' (Rüse 2014, p. 231). The literature does not suggest that they would form permanent voting coalitions in the Council, even if geographically close member states are more likely to vote similarly in the Council (Plechanovová 2011, Bailer *et al.* 2015). Instead, the literature stresses that the main reason why states engage in this kind of cooperation is linked to how the coalition can enhance the countries' bargaining position through the exchange of information, the pooling of expertise and their mutual support for a common legitimising strategy within the Council, i.e. rhetorical action (Rüse 2014).

I utilise an approach based on sociological new institutionalism. This does not mean that the approach would reject the possibility of the V4 acting as a territorially based coalition in its relations to the other members of the EU. It, however, suggests that the V4 could be considered as something more than merely a bargaining coalition despite its comparatively underdeveloped formal institutions (see the discussion below). The crucial assumption here is that the V4 contributes to the development of a shared logic of the appropriate thing to do among its members. The suggestion is that 'norms, values and routines embedded within institutions' (Bulmer 1998, p. 375) contribute to the development of a shared logic of the appropriate thing to do in a certain situation.

There are two different understandings of how institutional norms can shape the behaviour of states within an international organisation. The first is linked to the process of socialisation. This assumes that norms and values constitute the identity of the actors and thus shape their preferences (see, e.g. March and Olsen 1998, Finnemore and Sikkink 1998, Elgström 2005). Socialisation can then be defined as a process in which 'an agent switches from following a logic of consequences to a logic of appropriateness; this adoption is sustained over time and is quite independent from a particular structure of material incentives or sanctions' (Checkel 2007, p. 5–6). The second suggestion is that of rhetorical action developed by Frank Schimmelfennig (2001). This is a 'softer' interpretation of the role of norms, and the concept occupies a middle ground between rationalist and sociological institutionalism (Cf. Pollack 2009). Schimmelfennig assumes the 'strategic use of norm-based arguments' (2001, p. 76). However, the condition for this is that the argument is based on the actors' prior identification with the constitutive ideas of the community, which is then an argument that presupposes the existence of the logic of appropriateness.

Compared to the numerous studies devoted to the EU from a perspective based on sociological new institutionalism, surprisingly few efforts have been devoted to sub-regional groupings by scholars working in this tradition. A plausible explanation might be that the institutionalist literature within European integration tends to have an overly narrow focus on formal institutions (Suarugger 2016, p. 74). Yet, institutionalist theory within international relations has been developed based on a broad understanding of institutions going beyond formal institutions and being defined as constituted by 'persistent and connected sets of rules (formal and informal) that prescribe behavioural roles' (Keohane 1989, p. 3; Cf. Bulmer 1998).

Within EU studies, sociological new institutionalism is in particular often used as the theoretical base for studies of the top-down Europeanisation process (see, e.g. Radaelli 2003, p. 30). There is a tendency in the European integration literature to conflate socialisation with Europeanisation and to view it as a one-sided process of persuasion (Adler-Nissen 2014, p. 150) narrowly linked to the operations of various EU institutions.

The suggestion here is that the V4 in a similar way serves as a platform for socialisation. This is in line with an emerging, more complex view of socialisation within the Europeanisation literature, which reflects the complexity of the existence of various domestic norms that might or might not be compatible with EU-promoted norms (Cf. Tulmets 2015). The Visegrad cooperation might be supportive of the Europeanisation process but it can also reinforce alternative domestic norms compared to those promoted through the EU. However, it might be difficult in empirical terms to differentiate between socialisation and the strategic use of norms, i.e. rhetorical action. The rhetorical action idea speaks more to the relationship between the V4 and the other EU member states, i.e. the rhetoric of the V4 needs to refer to commonly shared norms and values in the EU in order to be successful in negotiations. However, the argumentation also needs to be in line with the internal ideas/values of the V4 community for this to be a viable strategy.

The Visegrad group and EU negotiations

The Visegrad Group itself claims on its website that the countries' cooperation ' … is not institutionalised in any manner' (Visegrad Group n.d.). This statement refers to a narrow interpretation of the meaning of the word 'institutionalised' (Fawn 2013). Even if the Visegrad Group does not have a permanent seat with a secretariat, it is a group with a rotating presidency, regular summits, meetings of other ministers, regular pre-meetings at different levels of the EU Council hierarchy, and extensive presidency programmes; it also has a Visegrad fund in place, which finances a large part of the research on the Visegrad cooperation (see, e.g. Kořan 2010, Beneš and Braun 2012).

In the literature, the V4 cooperation was initially categorised as a case of a complement/pre-accession instrument, thus suggesting that the main goal of the cooperation is 'to join a larger and more developed regional project, and sub-regional cooperation exists primarily to support and assist this process' (Dangerfield 2008, p. 633). There is, however, a consensus in the literature regarding the V4's continued relevance despite its having accomplished this task (see, e.g. Dangerfield 2008). Rick Fawn (2013) even argues that in ' … recent years Visegrad has become very evident in generating its own identity and promoting that within itself and outwards' (p. 340). According to Michal Kořan (2010) 'the dense interaction network contributes to socialisation and to the formation of a quasi-Visegrad identity' (p. 117). Crucial for the functioning of the V4 was the initiative of the Slovak V4 presidency in 2010–2011 to organise meetings of the V4 Prime Ministers ahead of every European Council meeting.

The V4 has often been understood as a grouping representing the entire central and eastern parts of the EU, but one reason for the success of the cooperation is likely to be that the countries have ruled out the possibility of enlarging the group to include more members. Instead meetings with representatives of other countries take the format of V4+ meetings. These are meetings where politicians or officials of other countries are invited to participate, often after an identification of shared interests between the invited countries and the V4 (Kořan 2010, p. 118). In the case of climate and energy policy such countries have in

the past included the Baltic States and, during the negotiations examined here, primarily Bulgaria and Romania. Thus, in both these cases,
the initial negotiations were between the V4 countries, and when it was
clear that the other countries had a similar position on the concrete
issue in question, they were invited to join the V4-led coalition on the
issue.

The Visegrad group and EU climate policy

The climate policy field is not defined as a key area of the V4. Yet, climate
issues are frequently referred to either in relation to the Visegrad Group's
energy cooperation, or because of developments regarding EU climate
policy (Törö et al. 2014). Moreover, energy cooperation has, despite the
countries' rather divergent preferences in the field, developed continuously
over the last 15 years. The Hungarian Visegrad presidency programme for
the 2001–2002 period already called for the need to coordinate the energy
policies of the Visegrad countries. Also, since the early 2000s a shared
priority of the V4 has been energy security. The group's efforts in this
regard increased in intensity after the Ukraine-Russian gas crisis in 2007
and 2009, even if the countries' different positions on Russian pipelines
caused some problems. The shared justification for the collaboration was
that the V4 countries viewed their position as vulnerable given their
relatively high reliance on natural gas imports from Russia (Nosko 2010,
Törö et al. 2014).

The group's initial interest in the climate policy field, prior to EU
accession, was largely determined by the demand to adapt to the EU's
environment and energy acquis (Törö et al. 2014). Their later cooperation
in the field has often been in response to developments at the EU level. The
earliest example of a successful Visegrad cooperation on EU climate policy
was linked to the climate and energy package negotiated in 2008. In
that year the countries did not immediately understand the potential costs
of the proposed package (Braun 2014b, p. 450), but when they did, they
came forth with a joint statement where they stressed that climate change
policy should be done in a 'fair way' (Joint Statement 2008). In the statement, the V4 framed the issue largely as an economic one, stating that the
EU's low income countries should be compensated for costs resulting from
the package. Later on, the three Baltic countries joined the Visegrad countries in a demand for the free distribution of some allowances for power
plants and a ceiling for the carbon price. They also called for a more
generous solidarity mechanism to help the poorer Member States in their
transition to a low-carbon economy (Braun 2014a, p. 121). Moreover, the
V4 and the three Baltic states then stood united in a call for consensual

decision making in the area and also called for 'efficient solidarity mechanisms' (Press Release of the Polish V4 Presidency).

In retrospect, the collaboration on the climate and energy package has largely been viewed as successful. The V4 managed to get the final decision on the package moved up from the Environment Council to the European Council, and the V4 received a satisfying solidarity mechanism (the burden sharing agreement) (Cf. Törö *et al.* 2014). Only occasionally have political parties in the region, such as the Polish Law and Justice Party (PiS) in 2011, challenged and called for a re-negotiation of the package (Braun 2014b, p. 144).

The V4 has since monitored the EU agenda in the field and at the regular meetings at the levels of the environmental attachés and the level of the ministers of environment, but also at meetings between the state secretaries for European affairs, the agenda has been discussed to see whether an issue is emerging on which the countries could find a common position.[2] As I show in the next part, the V4 joint positions' have since been coherent in their emphasis on cost reduction either in the sense of avoiding costly reforms and protecting the competitiveness of the countries' economies, or in demanding compensation for transitional costs.

The EU 2030 climate and energy framework

The V4's resistance to the EU's 2030 climate and energy framework received all of Europe's attention after the group's declaration of 30 September 2014, in which the group, among other things, rejected binding targets on renewable energy and binding energy efficiency targets at any level. The group also stressed the need for a realistic greenhouse gas reduction target, as well as the necessity of reaching the target in a 'technology neutral way'. More specific demands related, as the group's demands did in 2008, to the compensatory mechanism for poorer EU countries. Also in parallel to their demands regarding the 2008/9 package, the countries insisted on the importance of consensual decision making at the level of the European Council, and insisted that 'the European Council should, on [a] regular basis, review all important issues related to the 2030 climate and energy framework preparation and implementation, including possible impacts on competitiveness and energy prices' (Joint Statement 2014, p. 2).

The declaration of September 2014 was the result of a longer period of coordination between the countries on the issue, which went back to the negotiations on the climate and energy package in 2008 outlined above. The more specific work with the 2030 framework was initiated during the Polish V4 presidency in 2012–2013. The ministers of environment of the countries then agreed on a joint Concept Paper regarding the Commission's Green

Paper on the Climate and Energy 2030 Vision. Since Bulgaria and Romania also supported the position of the V4 they were invited to participate in the meeting where the paper was issued, and the cooperation then took the form of V4 + . The work took on an increased intensity after the Commission delivered its Communication on the framework in January 2014. From January 2014 until the decisive European Council meeting in October of the same year there were three meetings between the ministers of environment in addition to their regular pre-council meet-ings and other meetings at lower levels in the Council hierarchy.[3] In addition, the state secretaries for European affairs became the crucial level for coordinating the countries' positions.[4]

In the concept paper, the V4 called for flexibility in the 2030 framework, and argued that ' ... only tools acknowledging Member State differences are appropriate for satisfying properly energy and climate objectives' (Concept Paper on the Climate and Energy 2030 Vision). Furthermore the countries called for a defensive EU approach in the then upcoming global climate negotiations of 2015. In their view the EU's approach should respond to promises of other countries and actors rather than lead the development: 'Having a share of only about 11% of current global emissions, the EU cannot solve worldwide climate change alone.' (Concept Paper on the Climate and Energy 2030 Vision).

Despite the fact that the concept paper was negotiated by the countries' ministers of environment the paper hardly included any formulations about how climate change should be mitigated. The key words of the document were rather competitiveness and energy security, as this quotation illus-trates: 'the objectives of competitiveness and security of energy supply are not addressed fully in the current framework', and the paper also warned that the policy 'could negatively impact jobs and energy security, and lead to de-industrialisation' (Concept Paper on the Climate and Energy 2030 Vision).

More specifically the paper called for differentiated targets within the Emissions Trading System (ETS), and argued that opt-outs and differentia-tion can, 'by properly reflecting differentiated national circumstances[,] ... optimise costs to meet the objectives'. The paper stressed the importance of respecting the right of each EU member state enshrined in the Lisbon Treaty to shape their energy mix according to its own preferences. In this regard it emphasised the importance of 'indigenous sources of fossil fuels' for reducing the EU's energy prices as well as energy import dependence, and thus increasing the energy security of the Union.

The key components of the 2013 concept paper remained the main parts of the V4 joint position in the 2014 negotiations. These main points were also reflected in the Hungarian presidency program for the period 2013–2014, which in particular stressed that the priority should be increasing competitiveness and making energy prices more affordable, as well as avoiding carbon leakage. The Slovak program for 2014--2015 period, and thus for the time of the EU negotiations on the 2030 policy framework, again stressed the need for Visegrad coordination over the issue, arguing that each country should have the right to freely choose its most suitable energy mix, among other arguments. The program also called for an EU agreement to ensure fair effort sharing between member states and 'relevant mechanisms to compensate [the] excessive costs borne by [the] V4 countries' (Program of the Slovak Presidency in the Visegrad Group, 2014-2015). The program also elaborated on the potential costs of renewable energy support, and argued for the need to decrease the negative impacts on the energy costs of industry of support schemes for renewable energy .

The Commission presented its Communication on the issue in January 2014. In the Communication it was acknowledged that the consequences of the climate and energy policy would be disproportionally larger for EU member states with a lower Gross Domestic Income (A Policy Framework for Climate and Energy in the Period from 2020 to 2030). For this reason the V4 group demanded that the Commission investigate the consequences of the climate and energy policy on the member state level, which, however, did not happen (Joint Statement; Prouza 2014). Given the expected larger costs for low income countries, the Commission's position was that the compensation mechanisms agreed on in 2008 should remain in place, but that they should be reformed. The Commission proposed a new modernisation fund, which was to support energy and energy efficiency in the poorer member states, and also proposed that the existing innovation fund be reformed so that it would also include industrial innovation. Even if the V4 wanted to see a larger modernisation fund with the possibility of broader use, this was not a main issue of contestation during the negotiations. More controversial was the V4 demand for a different formula for calculating the distribution of allowances for low GDP countries. In particular it was crucial to the V4 group that only countries with a GDP below 60% of the average GDP in 2013 should be allowed to benefit from the mechanism (EurActiv.cz 2014).[5]

Moreover, the V4 group not only demanded the continuation of Article 10c of the ETS directive, which allowed most of the central and east European countries to give some free allowances to the energy sector, but also wanted this share of the allowances to be substantial. This, however, was primarily a demand from Poland, Hungary, Bulgaria and Romania. Czech representatives expressed their support, but did not consider the demand to be crucial (Geussová and Vítková 2014). Slovakia, on the other hand, was not entitled to the derogation during the 2013–2020 period, and so the issue was even less relevant for it (EurActiv.sk 2014). Yet, the country agreed to include the demand in the common position of the V4 (Program of the Slovak Presidency in the Visegrad Group).

Targets on the share of renewable energy were viewed as a problem because they could violate the member states' autonomy when it comes to deciding upon their own energy mix, and thus this could violate the provision of the Lisbon Treaty on technology neutrality. However, the countries weakened their resistance when it became clear that the targets on the share of renewable energy was a strong priority of other member states. A target level of energy efficiency was not included in the Commission's original proposal. The Czech Secretary of State for European Affairs described the change of the Czech position toward accepting non-binding targets as a way to avoid isolation at the EU level (Geussová and Vítková 2014).[6]

The Czech approach

The Czech position was prepared in the Ministry of Industry and Trade and the Ministry of Environment.[7] However, to allow for the development of a coherent national position the government's office was heavily involved in the preparation work, which was a condition for the coherency of the positions of the V4 countries.[8]

The initial Czech position was to favour one single target of a 35% reduction of greenhouse gas emissions compared to the levels of 1990. The country could, however, accept a target of 40% conditioned by the outcome of the Paris 2015 conference. The 35% target was viewed as more realistic, since this goal could be reached by the country while using the same measures as during the previous period. One of the government's main concerns was the quality of the new burden-sharing agreement, and how far this would provide satisfactory compensation for low-income countries such as the Czech Republic. The country was against having separate targets for renewable energy and energy efficiency. One strong argument here was that a separate renewable energy target could reduce the price of allowances within the EU ETS system, which would then weaken the effects of the ETS (Rámcová pozice[9]).[10] However, there was also a concern that the renewable energy target could violate the principle of each member state's sovereignty to decide about its energy mix and that the energy efficiency target could be economically costly for the country (Rámcová pozice).

Most of these demands were thus in line with the V4 position. However, there were some issues where the Czech position slightly departed from the V4 position. The Czech Republic, for instance, supported the opinion of the Commission that the emission target should be adopted in 2014 and not during the Paris 2015 negotiation as the V4 2013 concept paper had declared. This was also in contradiction to the position of Poland and Hungary (Rámcová pozice; Concept Paper on the Climate and Energy 2030 Vision).

However, the Czech position was in line with V4 agreements on several crucial priorities: the call for one single target, the insistence on the continuation of the solidarity mechanism, and the emphasis on taking into account the effects on the competitiveness of the low-income economies. The Czech Republic, in its position paper, also warned of a climate and energy policy that would give disproportionate priority to the climate goal and called for a 'return to the basic concepts of the Union energy policy, i.e. a balance of all three of the main strategic goals: sustainability, security of supply and competitiveness' (Rámcová pozice – my translation).

The main priority, however, was to make sure that decisions regarding the framework and future changes of this policy remained consensual and within the European Council. As State Secretary Tomáš Prouza stated, 'the condition of the success of the cooperation was that we managed to stick to our position and get the issue all the way up to the European Council.'[11] Also, other Czech representatives argued that there was a genuine concern that, if qualified majority voting had been applied, some proposals could have been devastating for the countries' economies.[12] In addition, this feeling of being under threat made it easier to find agreements with the other V4 countries.

Two of the interviewed Czech representatives, however, stressed that there was one major underlying difference in the positions of the Czech Republic and Poland. The Czech position was based on accepting the intention of the package but focused on getting, compensation as high as possible. The Polish position, on the other hand, was focused on avoiding changes in the country's economic structure.[13] An example of these divergent strategies was the negotiations on the Market Stability Reserve in April 2015. An attempt to form a blocking minority by Poland and Hungary, among others, failed after the Czech Republic gave up on its resistance towards the idea. The fourth Visegrad country, Slovakia, did not participate in the blocking coalition at any time during the negotiations. The Czech Republic's changed position allowed for the introduction of the new mechanism in 2019 instead of the originally planned 2021 (Czech News Agency 2015). The Czech Republic initially tried to maintain a joint position with Poland. However, in the end the Czech side wanted a deal because a price increase of ETS allowances would also increase the available means in the modernisation fund, from which the country could benefit. Poland, on the other hand, was concerned that a decrease in ETS allowances would also affect the volume of allowances that could be distributed without auctioning in accordance with derogation (Article 10c of the ETS Directive).[14]

If we look to the final agreement from October 2014 we can see that the coalition was successful in getting through several of the Czech Republic's main demands – particularly the agreement that '[t]he European Council will keep all the elements of the framework under review ... ' (Conclusions on the 2030 Climate and Energy Policy Framework) and the agreement on the compensatory mechanism, where the formula for calculating which countries should be allowed to benefit from the compensatory mechanism was designed in accordance with the preferences of the V4 countries. In other words, the criterion that only countries with GDPs below 60% of the average GDP in 2013 would be included in the mechanism was beneficial to all the countries of the coalition, but by a narrow margin excluded Greece from the mechanism (see Mitafidis *et al.* 2016 for a critical reflection on this). Moreover, the V4+ demand for the extension of the derogation on compulsory auctioning for the energy producing sector (Article 10c of the ETS Directive) was also satisfied. This, however, was a less crucial demand for the Czech Republic and Slovakia than for Poland (Geussová and Vítková 2014).

Thus, if we were to assess the benefits for the Czech Republic of being part of a V4 coalition on this issue, the judgement would most likely be positive. The Czech Republic, however, was more willing to compromise on some issues than the other V4 countries: the country was faster in accepting a change in its position and accepting non-binding targets at the national level for energy efficiency and the share of renewable energy (Denková 2014). Another added value of the cooperation was that the Visegrad Group, as a consequence, became better known to other EU countries and widely viewed as a counterweight to the Green Growth Group.[15] An indication of this was the French and German interest in participating in the final V4 meeting prior to the October 2014 European Council.[16]

The logic of appropriateness, rhetorical action and socialisation

As the above overview of the V4's approach to EU climate policy indicates, the primary goal of the countries can be summarised as the avoidance of costs and, more importantly, to get compensation for any reforms that might bring about costs. The V4 group's demands have largely been developed in opposition to the EU tradition of framing climate policy in the vocabulary of ecological modernisation, a principle that suggests that stringent environmental rules can be combined with and even boost economic growth (Wurzel and Connelly 2011).[17] According to Selin and VanDeveer (2015) 'ecological modernisation has come close to being an official ideology of the EU' (p. 47). The EU's climate policy has been promoted based on the logic of ecological modernisation and the argumentation that the climate mitigation policy would not only be instrumental in decreasing global warming, but would also 'create thousands of new businesses and millions of jobs in Europe' (José Manuel Barroso quoted in Jordan and Rayner 2011, p. 75). Moreover, the key component of the climate policy, the ETS, is a market-based instrument in line with the ideas of the principle (Bailey *et al.* 2011, p. 686, Braun 2011a).

Rhetorical action

The V4 took the position that the transition to a low carbon economy was a challenge to economic growth, and argued that the EU's climate policy hampered the region's global competitiveness. The countries articulated a shared understanding of themselves as being in a specific vulnerable position compared to the EU member states in Western Europe. According to them, a transition to a less carbon intense economy by necessity would bring them costs instead of opportunities. The argumentation fulfils the criteria of rhetorical action if it follows in line with and appeals to 'constitutive ideas of the community' (Schimmelfennig 2001, p. 77).

The V4 argumentation can be described as rhetorical action for two reasons. First, the V4 countries' joint emphasis on the issue of fairness referred to a value of solidarity assumed to be shared between all the countries of the EU and well-established within the EU's cohesion policy (Cf. Bachtler *et al.* 2013).

Second, even if the V4 argumentation contradicted the basic assumption of ecological modernisation that economic growth and climate change mitigation are combinable targets, the argumentation still followed the same underlying line of economic reasoning. The V4 countries' concern that their economies could be the ones bearing the actual cost of the reforms through decreased competitiveness is in line with the predictions steaming from an ecological modernisation perspective. As John Bellamy Foster (2012) has pointed out, ecological modernisation theory is a continuation of modernisation theory, where environmental conflicts are understood as pathologies that will be transcended by modernisation itself (Foster 2012), and therefore, according to this theory, the west European countries are closer to transcending the conflict than countries less developed in economic terms. For this reason, it is not a coincidence that the principle of ecological modernisation has developed primarily in western Europe (Mol 2003), and it is thus in accordance with this line of reasoning that low-income economies would be sceptical about the principle.

Socialisation

There have been suggestions in the socialisation literature that the differences between more recent EU member states and more long-term members would gradually decrease over time (see, e.g. Beyers 2010). However, the V4 approach to the EU's climate policy suggests the opposite and a high degree of continuity in the group's approach as discussed above. The strategy of cost reduction can even be viewed as a continuation of the countries' approaches to EU environmental policy prior to and after their accession to the EU in 2004 (Braun 2014b). This suggests the limitations of socialisation based on Europeanisation, because from such a perspective we would have expected increasing support for the EU approach based on ecological modernisation, which has not been the case.

In the Czech case, we can even find indications of the opposite development. During the negotiations on the EU Climate and Energy Package in 2008 there were some strong voices within the political elite supporting a more progressive climate policy in line with the ideas of ecological modernisation.[18] As a consequence there was a contestation within the political elite over the issue which manifested itself, among others, in internal conflicts between the country's Ministry of Environment and Ministry of Industry and Trade (see Braun 2014b, p. 127).[19] This domestic contestation over EU climate politics also led to some tensions between the country and the other V4 states in the 2008 negotiations.[20] In 2014, in contrast, there was more consensual support within the Czech political elite for the joint V4 approach challenging the proposed EU climate targets and primarily focusing on cost reduction for the country.[21]

Even if this does not necessarily suggest the existence of an alternative V4 process of socialisation, the Visegrad Group's coherent approach on the issue and coordination of the issue between the V4 countries with a strong involvement of the government's office[22] is likely to have reinforced the view in the country that an ambitious EU climate mitigation policy would by necessity have a negative impact on the Czech economy and excluded alternative interpretations. Such an interpretation would be in line with findings from the Czech think tank literature on the V4 group as an increasingly central point of reference for the politicians and civil servants who are formulating the country's position in EU negotiations (Cf. Kořan 2010, Dostál 2015).

Conclusion

The Visegrad Group was successful in defending its positions in the negotiations on the EU 2030 climate and energy framework. The success of the Visegrad Group on this issue is a further indication that the scholarly community should take this form of cooperation seriously. I suggest that even if the V4 can be described as a successful territorially based coalition within the Council negotiations, such a description ignores the more complex internal dimension of this institutionalised form of cooperation. The close interaction between the state representatives involved suggests that scholars should take seriously the possibility that a shared view of appropriate behaviour is developing within the V4. The group's emphasis on economic solidarity and compensation for low-income economies is an argument that corresponds well with the shared self-understanding of the countries as being in a specific vulnerable position compared to the EU member states in western Europe. The close interactions between the countries are likely to reinforce such a defensive interpretation of the consequences of climate mitigation policy on them. Therefore, the V4 should be considered a potential platform for socialisation despite its relatively weakly developed formal institutions. However, the V4's emphasis on economic solidarity can also be interpreted as an example of rhetorical action. From such a perspective it was strategic of the V4 to focus on economic compensation given that such an argument corresponds to the shared EU norm of economic solidarity. These two interpretations are not mutually exclusive. The condition for the rhetorical action to work is that the argumentation has to be in line with an understanding of appropriateness that is acceptable in both settings, i.e. in the Visegrad Group and also in the EU.

In conclusion, it should be stressed that the analysis I have presented here should be viewed as an early contribution to the study of sub-regional cooperation within the EU. There has been a tendency among both scholars and practitioners to neglect the importance of institutionalised cooperation in European sub-regions. I suggest that these institutions matter and should be the subject of further examination.

Notes

1. The interviewed officials are kept anonymous with the exception of Tomaš Prouza – the Secretary of State during the relevant period, who agreed to be quoted by name. Most of the interview material referred to was gathered in 2017, some references are also made to a previous set of interviews from 2009 and 2011.
2. Author's interview, 24 February 2017, Ministry of Environment, Prague.
3. Author's interview, 17 February 2017, Ministry of Environment, Prague.
4. Author's interview, 23 March 2017, Office of the Government, Prague.
5. 24 February 2017, Ministry of Environment, Prague.
6. Author's interview, 23 March 2017, Office of the Government.
7. Author's interview, 24 February 2017, Ministry of Environment, Prague.
8. Author's interviews, 24 February 2017, Ministry of Environment, Prague; 23 March 2017, Office of the Government.
9. Framework position (author's translation). The reference refers to the Czech government's framework position presented to the Czech parliament.
10. Author's interview, 23 March 2017, Office of the Government.
11. Author's interview, 23 March, Office of the Government – my translation.

12. Author's interview, 24 February 2017, Ministry of Environment, Prague.
13. Author's interview, 24 February 2017, Ministry of Environment, Prague; Author's interview, 23 March, Office of the Government.
14. Author's interview, 24 February 2017, Ministry of Environment, Prague.
15. A group initiated by the UK, and including Belgium, Denmark, Estonia, Finland, France, Germany, Italy, the Netherlands, Portugal, Slovenia, Spain and Sweden, it demands more ambitious EU climate change polices (Wurzel *et al.* 2017).
16. Author's interview, 24 February 2017, Ministry of Environment, Prague.
17. Author's interview, 17 February 2017, Ministry of Environment, Prague.
18. Mainly but not exclusively ministers and members of parliament from the Green Party, which in 2008 was a junior coalition partner in the government but has since become largely irrelevant in Czech politics (see, e.g. Frankland 2016).
19. Author's interviews, 11 February; 6 August 2009, Ministry of Environment, Prague; 17 August 2011, Ministry of Industry and Trade, Prague.
20. The country, for instance, did not support the demand that 1990 be used as the base year for calculating the carbon reduction targets as proposed by Poland and Hungary (Author's interviews, 17 August 2011, Ministry of Industry and Trade, Prague; Braun 2014a, p. 450).
21. Author's interviews, 23 March 2017, Office of the Government, Prague.
22. Author's interview, 23 March 2017, Office of the Government, Prague.

Disclosure statement

No potential conflict of interest was reported by the author.

Funding

This work was supported by Metropolitan University Prague under research scheme no. VVZ 34-04.

ORCID

Mats Braun ⓘD http://orcid.org/0000-0001-7574-3253

References

A Policy Framework for Climate and Energy in the Period from 2020 to 2030, COM (2014)15, final.
Adler-Nissen, R., 2014. Stigma management in international relations. *International Organization*, 68 (1), 143–176. doi:10.1017/S0020818313000337
Bachtler, J.C., Mendez, F., and Wishlade, 2013. *EU cohesion policy and european integration: the dynamics of EU budget and regional policy reform*. Farnham: Ashgate.
Bailer, S., Mattila, M., and Schneider, G., 2015. Money makes the EU go round. The objective foundations of conflict in the council of ministers, in. *Journal of Common Market Studies*, 23 (3), 437–456. doi:10.1111/jcms.12194

Bailey, I., Gouldson, A., and Newell, P., 2011. Ecological modernisation and the governance of carbon: a critical analysis. *Antipode*, 43 (3), 682–703. doi:10.1111/j.1467-8330.2011.00880.x

Beneš, V. and Braun, M., 2012. Evropský rozměr české zahraniční politiky. *In*: M. Kořan, ed. *Česká zahraniční politika v roce 2011: analýza ÚMV*. Prague: Ústav mezinárodních vztahů, 67–94.

Beyers, J., 2010. Conceptual and methodological challenges in the study of European socialization. *Journal of European Public Policy*, 17 (6), 909–920. doi:10.1080/13501763.2010.487004

Braun, M., 2014a. EU climate norms in East-Central Europe. *Journal of Common Market Studies*, 52 (3), 445–460. doi:10.1111/jcms.12101

Braun, M., 2014b. *Europeanization of environmental policy in the new Europe: beyond conditionality*. Farnham: Ashgate.

Bulmer, S.J., 1998. New institutionalism and the governance of the single European market. *Journal of European Public Policy*, 5 (3), 365–386. doi:10.1080/135017698343875

Checkel, J.T., 2007. *International Institutions and Socialization in Europe - Introduction and Framework*, 3–28. Cambridge: Cambridge University Press.

Concept Paper on the Climate and Energy 2030 Vision. *In*: 19th Meeting of the Environment Ministers of the Visegrad Group Countries, April 2013 Bulgaria and Romania.

Conclusions on the 2030 Climate and Energy Policy Framework, European Council (23 and 24 October 2014).

Cottey, A., 2009. *Sub-regional cooperation in Europe: an assessment*. Bruges: College of Europe.

Czech News Agency, 2015. *Země EU se předběžně dohodly na reformě ETS od roku 2019* [online]. Available from: http://zpravy.aktualne.cz/ekonomika/zeme-eu-se-predbezne-dohodly-na-reforme-ets-od-roku-2019/r~5d077324ee9311e497be0025900fea04 [Accessed 17 March 2017

Dangerfield, M., 2008. The Visegrád group in the expanded European Union: from preaccession to postaccession cooperation. *European Politics and Societies*, 22 (3), 630–667. doi:10.1177/0888325408315840

Darby, M., 2014. *Poland "Won" EU 2030 Deal – does the climate lose?* [online] Climate Home. Available from: http://www.climatechangenews.com/2014/10/29/poland-won-eu-2030-deal-does-the-climate-lose/ [Accessed 17 March 2017

Denková, A., 2014. *PŘEHLED POZIC – ČR a Střední Evropa: energetickou politiku si chceme vybrat sami*. EurActiv. https://euractiv.cz/section/aktualne-v-eu/news/prehled-pozic-cesko-a-stredni-evropa-o-energeticke-politice-si-tentokrat-chceme-rozhodnout-sami-012196/

Dostál, V., 2015. *Trends of Visegrad foreign policy*. Prague: Association for International Affairs (AMO).

Elgström, O., 2005. Consolidating 'Unobjectionable' norms: negotiating norm spread in the European Union. *In*: O. Elgström and C.H. Jönsson, eds. *European union negotiations processes, networks and institutions*. London: Routledge, 29–45.

EurActiv.cz, 2014. *Klimaticko-energetické cíle 2030 a pozice ČR* [online]. Available from: http://euractiv.cz/factsheet/energetika/klimaticko-energeticke-cile-2030-a-pozice-cr-000104/[Accessed 17 March 2017

EurActiv.sk, 2014. *Klimatická a energetická politika EÚ do roku 2030* [online]. Available from: https://euractiv.sk/fokus/energetika/klimaticka-a-energeticka-politika-eu-s-vyhladom-do-roku-2030-000333/.

Fawn, R., 2013. Visegrad: fit for Purpose? *Communist and Post-Communist Studies*, 46 (3), 339–349. doi:10.1016/j.postcomstud.2013.06.004

Finnemore, M. and Sikkink, K., 1998. International norm dynamics and political change. *International Organization*, 52 (4), 887–917. doi:10.1162/002081898550789

Foster, J.B., 2012. The planetary rift and the new human exemptionalism: a political-economic critique of ecological modernization theory. *Organization & Environment*, 25 (3), 211–237. doi:10.1177/1086026612459964

Frankland, G., 2016. Central and Eastern European green parties: rise, fall and revival?. *In*: E. van Heute, ed.. *Green parties in Europe*. London: Routledge, 59–91.

Geussová, M. and Vítková, E., 2014. Interview with T. Prouza - Nemá smysl hrát si na husity. *Pro-Energy Magazín* [online]. Available From: https://www.vlada.cz/assets/evropske-zalezitosti/aktualne/pro-energy-rozhovor.pdf [Accessed 17 March 2017].

Joint Statement of the 15th Meeting of Ministers of Environment of the Visegrad Group Countries, 18–19 September 2008, Budapest, Hungary.

Joint Statement of the 21st Meeting of the Ministers of Environment of the Visegrad Group Countries, the Republic of Bulgaria and Romania, 30 September 2014, Bratislava, Slovakia.

Jordan, A. and Rayner, T., 2011. The evolution of climate policy in the European Union: an Historical overview. *In*: A. Jordan, *et al.*, eds.. *Climate change policy in the European Union: confronting the dilemmas of mitigationand adaptation?* Cambridge: Cambridge University Press, 52–80.

Keohane, R., 1989. Neoliberal institutionalism: a perspective on world politics. *In*: R. Keohane, ed. *International institutions and state power*. Boulder, CO: Westview Press, 1–20.

Kořan, M., 2010. The Visegrad cooperation, Poland, Slovakia, and Austria in the Czech foreign policy. *In*: M. Kořan, ed. *Czech foreign policy in 2007-2009: analysis.* . Prague: Institute of International Relations, 115–147.

March, J.G. and Olsen, J.P., 1998. The institutional dynamics of international political orders. *International Organization*, 52 (4), 943–969. doi:10.1162/002081898550699

Mitafidis, A., Chatzianastasiou, C., and Baxevanis, P., 2016, Including Greece in ETS compensation mechanisms. *EurActive* [online]. Available from: https://www.euractiv.com/section/energy/opinion/including-greece-in-eu-ets-compensation-mechanisms [Accessed 17 March 2017

Mol, A.P.J., 2003. The environmental transformation of the modern order. *In*: T. J. Misa, P. Brey, and A. Feenberg, eds. *Modernity and technology*. Cambridge: MIT Press, 303–326.

Nosko, A., 2010. Regional energy security: Visegrad finally at Work? *In*: M. Majer, *et al.*, eds.. *Panorama of global security environment*. Bratislava: CENAA, 79–91.

Plechanovová, B., 2011. The EU council enlarged: north–south–east or core–periphery? *European Union Politics*, 12 (1), 87–106. doi:10.1177/1465116510390720

Pollack, M., 2009. The new institutionalisms and European integration. *In*: Wiener and Diez, eds. *European integration theory*. 2nd ed. Hampshire: Oxford, 125–144.

Press release of the polish V4 presidency after the official summit of the prime ministers of the Visegrad group countries and Baltic States. Available from: http://www.visegradgroup.eu/2008/press-release-of-the [Accessed 2 August 2017

Program of the Slovak Presidency in the Visegrad Group (July 2014–June 2015), Dynamic Visegrad for Europe and Beyond.

Prouza, T., 2014. *Opomenutá energetická bezpečnost* [online]. Available From: http://www.prouza.cz/blog/opomenuta-energeticka-bezpecnost/ [Accessed 17 March 2017].

Prouza, T., 2015. Interview in Wyszehrad nie umarł. *Gazeta Wzborcza* [online]. Available from: http://wyborcza.pl/1,75248,18034400,Wyszehrad_nie_umarl.html [Accessed 17 March 2017

Radaelli, C., 2003. The Europeanization of public policy. *In*: K. Featherstone and C. Radaelli, eds.. *The politics of Europeanization*. Oxford: University Press, 27–56.

Rámcová pozice/Stanovisko pro Parlament ČR: Sdělení Komise Evropskému parlamentu, Radě, Evropskému hospodářskému a sociálnímu výboru a Výboru regionů. Rámec politiky v oblasti klimatu a energetiky v období 2020-2030.

Rüse, I., 2014. Nordic-baltic interaction in European Union negotiations: taking advantage of institutionalized cooperation. *Baltic Studies*, 45 (2), 229–246. doi:10.1080/01629778.2013.846928

Schimmelfennig, F., 2001. The community trap: liberal norms, rhetorical action, and the Eastern enlargement of the European Union. *International Organization*, 55 (1), 47–80. doi:10.1162/002081801551414

Selin, H. and VanDeveer, S.D., 2015. *European Union and environmental governance*. London: Routledge.

Skjærseth, J.B., 2016. Linking EU climate and energy policies: policy-making, implementation and reform. *International Environmental Agreements: Politics, Law and Economics*, 16 (4), 509–523. doi:10.1007/s10784-014-9262-5

Skovgaard, J., 2013. The limits of entrapment: the negotiations on EU reduction targets, 2007–11. *JCMS: Journal of Common Market Studies*, 51 (6), 1141–1157. doi:10.1111/jcms.2013.51.issue-6

Suarugger, S., 2016. Sociological Approaches to the European Union in Times of Turmoil. *Journal of Common Market Studies*, 54 (1), 70–86. doi:10.1111/jcms.12330

Tallberg, J., 2010. Explaining the institutional foundations of European Union negotiations. *Journal of European Public Policy*, 17 (5), 633–647. doi:10.1080/13501761003748559

Törö, C., Butler, E., and Grúber, K., 2014. Visegrád: the evolving pattern of coordination and partnership after EU enlargement. *Europe-Asia Studies*, 66 (3), 364–393. doi:10.1080/09668136.2013.855392

Tulmets, E., 2015. *Norm transfer and circulation in international relations. the case of the European Union and its neighbourhood*, Thesis (habilitation), Sciences Po Paris.

Visegrad Group, n.d. *Aims and structure* [online]. Available from: http://www.visegradgroup.eu/about/aims-and-structure [Accessed 17 March 2017

Wurzel, R.K.W., Liefferink, D., and Connelly, J., 2017. Introduction: European Union climate leadership. *In*: R.K.W. Wurzel, J. Connelly, and D. Liefferink, eds.. *The European Union in international climate change politics: still taking a lead?* London: Routledge, 3–19.

Wurzel, R.K.W. and Connelly, J., 2011. Conclusion: the European Union's leadership role in international climate change politics reassessed. same authors, ed. *The European Union's leadership role in international climate change politics reassessed*. London: Routledge, 271–290.

Climate laws in small European states: symbolic legislation and limits of diffusion in Ireland and Finland

Diarmuid Torney ⓘ

ABSTRACT

The past decade has seen the introduction of framework climate change laws in several countries. The development of climate laws in two small European states, Ireland and Finland, both of which introduced national climate laws in 2015, are examined. Two questions are addressed. First, to what extent do later adopters of climate policy instruments draw on the examples of pioneering legislation? Second, how and why are pioneering climate policy instruments modified by later adopters? In both cases, the 2008 UK Climate Change Act was a source of inspiration in the early stages, particularly for civil society campaigns. Thereafter, domestic interests mobilised to remove from legislative proposals the most pioneering and ambitious parts of the UK model. The result, in both cases, was enactment of climate laws that resembled only very loosely the UK Climate Change Act.

Introduction

Recent years have seen a proliferation of legislation relating to climate change at national level. According to the Global Climate Legislation Study, as of April 2018 there were more than 1,500 climate change laws and policies worldwide, an increase from 72 at the time of the Kyoto Protocol in 1997 (GLOBE 2018). One noteworthy category of policy instrument is the set of over-arching framework climate change laws. The 2008 UK Climate Change Act (CCA) is an example of this legislative approach, and was the first of its kind. More recently, a range of other countries have introduced similar legislation.

The UK CCA was pioneering in three major ways. First, the CCA established a long-term unilateral greenhouse gas (GHG) emission reduction goal of at least 80% below 1990 levels by 2050. Second, it required the government to set out successive five-year 'carbon budgets' that set a maximum total emission limit for each five-year period that are legally binding. In principle, this opens the door to a legal action against the government. Third, the Act established an independent Committee on Climate Change, the role of which is to provide recommendations and advice to the government on carbon budgets and targets, as well as on climate policies. The Committee reports progress in achieving climate targets to Parliament on an annual basis. While the Committee's recommendations with respect to carbon budgets are advisory rather than binding, the government must provide an explanatory statement if it wishes to pursue a course of action other than that recommended by the Committee.

The development and enactment of the UK CCA benefited from an unusual constellation of supportive factors. These included a high level of international public attention to the climate issue, competition among political parties to 'out-green' each other, strong political leadership by leading Labour politicians Tony Blair, David and Ed Miliband, and Conservative leader David Cameron, and support from the business sector, encouraged by the Stern Review (Carter and Jacobs 2014). Civil society mobilized around a 'Big Ask' campaign led by Friends of the Earth (FOE) UK, which was instrumental in driving the legislative process (Carter and Childs 2018). These were indeed highly favourable – and unusual – circumstances that enabled the passing of a particularly pioneering piece of legislation. Subsequent circumstances in the UK differ profoundly, leading some to call into question the resilience of the CCA model in these radically altered circumstances (Lockwood 2013).

Here, I am concerned with whether later adopters draw on the examples of pioneering climate policy instruments and, in particular, what happens to such instruments when they diffuse to other jurisdictions. I follow the process of adoption of climate laws in two small European states, Ireland and Finland. Both adopted climate change laws in 2015: the Climate Action and Low Carbon Development Act in Ireland, and the Climate Change Act in Finland. These were both enacted some years after the UK CCA came into force and, in each case, following a long process towards adoption. Drawing also on the literature on incomplete policy diffusion and symbolic environmental legislation, these cases are used to answer two questions. First, to what extent do later adopters of climate policy instruments draw on the examples of pioneering legislation? Second, how and why are pioneering climate policy instruments modified by later adopters?

Ireland and Finland are chosen not just because both have recently adopted climate laws, but also because they share important characteristics. Both are small states and members of the European Union (EU). They also share arguably dubious climate credentials. Ireland's per capita emissions (13.5 tCO_2e) are third-highest in the EU and significantly higher than the EU-28 average of 8.7 tonnes (Eurostat 2018). Ireland will miss its EU 2020 emissions target for the non-ETS sector by a wide margin (EPA 2018). The Germanwatch/Climate Action Network Climate Change Performance Index (CCPI) for 2019 ranked Ireland 48th out of 60 countries, the lowest ranking of any European state (Burck *et al.* 2018). Finland's profile is more complicated. It was ranked 13th out of 60 in the 2019 CCPI rankings, but is sometimes considered a laggard among Nordic states. Its emissions per capita are the highest of all the Nordic states and, significantly, have been on an upward trajectory since 1990, whereas other Nordic countries' emissions have travelled in the other direction (Gronow and Ylä-Anttila 2016). Indeed, Finland has even been described as a 'failing ecostate' (Koch and Fritz 2014, p. 691).

My contribution here is to provide evidence from new case studies to examine propositions drawn from the literature on incomplete policy diffusion and symbolic legislation (Newig 2007, Klinger-Vidra and Schleifer 2014). I also examine in detail the behaviour of later adopters of innovative policy responses (Liefferink and Wurzel 2017). I use primary empirical research, including seven interviews conducted in May–June 2016 for the Irish case study and ten interviews conducted in August 2016 for the Finnish case study (see Appendix I for details). Interviewees were selected to be representative of the key stakeholders involved in each legislative process. The interviews were not recorded, but notes were taken and expanded upon immediately after each interview. Anonymity was promised to interviewees in all cases.

The next section combines elements of the literature on policy diffusion and on symbolic environmental legislation to develop a framework to study the (incomplete) diffusion of climate laws. The following two sections provide an analytical narrative of the pathway towards enactment of climate laws in Ireland and Finland. The final section discusses the findings of the cases in light of the questions posed above.

Diffusion and symbolic politics

Policy diffusion can be defined as 'when government policy decisions in one country are systematically conditioned by prior policy choices made in another country' (Simmons *et al.* 2006, p. 787). Recent research on the enactment of national climate change laws and policies supports the contention that policy diffusion is a driver of national policy development. Fankhauser *et al.* (2016, p. 319) argue that '[t]he propensity to pass climate legislation increases with the number of climate laws passed elsewhere'. They identify a need for further research to unpack the effect of diffusion, including the drivers of diffusion. However, large-N quantitative studies of policy diffusion cannot tell us which parts of a law or policy were diffused, whether the resulting policy was more or less stringent than the original source, and what explains variation between source and destination.

A recent strand of the diffusion literature emphasises that we should not necessarily expect full convergence between source and destination. As Klinger-Vidra and Schleifer argue, 'as they diffuse, norms, ideas, and practices often change in form and content' (Klinger-Vidra and Schleifer 2014, p. 264). Differing domestic political conditions may lead to less than full convergence across jurisdictions. Moreover, because policymakers in the recipient jurisdiction are likely to consider carefully the policy in question and to adopt only those parts they believe to be most appropriate to their national circumstances while discarding others, diffusion caused by learning from other jurisdictions may lead to incomplete convergence.

Incomplete convergence may be driven by incentives that subsequent adopters of policy innovations may have to adopt in limited or symbolic ways models of best practice from other jurisdictions. Newig (2007) defines symbolic environmental legislation as 'laws which despite their often ambitious officially declared objectives are designed to remain ecologically ineffective' that can be viewed as 'instruments[s] for *managing* rather than *resolving* environmental problems' (Newig 2007, p. 277). Distinguishing between *substantive* effectiveness – the ability of a law to effectively respond to an environmental problem – and *political-strategic* effectiveness – the ability of a law to serve instrumental ends such as responding to political pressures – Newig suggests that symbolic legislation is characterised by low substantive effectiveness and high political-strategic effectiveness. He identifies four conditions under which symbolic environmental legislation is likely to be adopted: first, where there is significant political pressure to respond; second, where there are no clear solutions or where short run costs outweigh short run benefits; third, where there is significant societal conflict; and fourth, where the issue is characterised by high levels of complexity. Building on the discussion of incomplete policy convergence above, these suggest conditions under which the recipients of policy diffusion may modify and downscale pioneering legislative models.

Climate change policy can be particularly conducive to symbolic action, because of the combination of high complexity and significant international pressure to act. Previous research has highlighted the ways in which states seek to use their responses to climate change to enhance their status both domestically and internationally. Vogler points to states' concern for recognition and prestige as a motivation for taking action on climate change (Vogler 2016), echoing Falkner's argument that the normative structure of international society has evolved to include a norm of global environmental responsibility (Falkner 2012). Similarly, Cass (2009) argues that states use environmental foreign policy to signal their adherence to international norms, and to manage international identities in the eyes of both external and domestic audiences. Climate change policy is also frequently charac-terised by high levels of societal conflict as well as by high short term economic costs and low short term economic benefits (Giddens 2011). Societal conflict can manifest itself in various ways, including through lobbying by economic interest groups (Newell and Paterson 2010) and through political party competition (Jensen and Spoon 2011, Little 2017). The case studies below will examine two propositions: first, that interna-tional pressure to act on climate change pushes policymakers to develop climate policies that draw on experiences from other jurisdictions; and second, that interest group mobilisation and political competition over climate policy will cause states to adopt weaker, more symbolic ver-sions of pioneering policies in other jurisdictions.

Diffusion of climate laws offers an opportunity to study how these dynamics play out, with the UK CCA serving as a potential source of (incomplete) diffusion. Because of its innovativeness and level of ambition, the UK CCA provides both a potential focal point for subsequent policy diffusion but also, precisely because of its level of ambition, a case where we are likely to see later adopters adopting more symbolic climate laws, particularly in those jurisdictions where the unusual circumstances that characterised UK climate politics in the late 2000s do not hold. In a similar vein, Benson and Lorenzoni (2014) argue that contextual con-straints may limit the applicability of the UK CCA to other countries. The next two sections discuss the passage to enactment of the Finnish and Irish climate laws, both enacted in 2015.

The Irish climate action and low carbon development act[1]

Ireland's climate law, eventually signed into law in December 2015 as the Climate Action and Low Carbon Development Act, was eight years in the making. The origins of the Act can be traced to the pan-European 'Big Ask' campaign run by FOE in many EU member states. In Ireland, this cam-paign was launched by FOE Ireland in June 2007 (FOE Ireland 2007). The centrepiece of this campaign was a call for a climate law for Ireland to contain a GHG reduction target of 60% by 2050 along with an annual GHG reduction target of 3%.

A general election in June 2007 resulted in a coalition between the centre-right Fianna Fail and Progressive Democrats parties and the Green Party, the first time the Green Party had entered government. Although the Green Party election manifesto had not included a specific commitment to enacting a climate change law, it did commit to 'seek an all-party approach to cut Ireland's carbon emissions by 3% annually' (Green Party 2007, p. 6). The programme for government agreed between the three coalition part-ners included this 3% annual reduction pledge as well as a commitment to producing an annual report on progress in meeting climate change targets. It also committed to establishing a 'Commission on Climate Change'.

The Green Party was allocated the environment and energy ministries, and made enactment of a climate law a central priority of their period in government. Over time, the UK CCA became a key reference point, particularly once the UK Act became law in November 2008. Green Party minister for energy Eamon Ryan stated that 'My preference would be to follow the British model for the Climate Change Act. The advantage is that we can benefit from legislation that is tried and tested and we can adapt it and make it better' (McGee 2009). The UK embassy in Dublin sought, as part of a broader climate diplomacy strategy, to promote the CCA as a model for climate governance. Its efforts included holding a workshop on the topic in Dublin, and the British Ambassador met with environment minister and Green Party leader John Gormley (Interview 3). Furthermore, the recently formed Stop Climate Chaos coalition of environmental and development NGOs increasingly pointed to the UK Act as a model that Ireland could follow as part of their campaigning (Interview 2).

The opposition Labour Party became active on the question of a climate law for Ireland during this period and similarly looked to the UK as well as further afield. Labour Party senator Ivana Bacik introduced a Climate Protection Bill in September 2007 (Houses of the Oireachtas 2007), which was broadly modelled on the Friends of the Earth proposal from earlier that year. It proposed a carbon budget for the period 2010–2050 and would establish a 'Commission on Climate Change'. Labour Party TD (MP) Liz McManus was appointed rapporteur of an Oireachtas (parliament) Joint Committee on Climate Change and Energy Security that published a proposal for a Climate Change Bill in October 2010 that was modelled quite explicitly on the UK CCA, including a 2050 GHG reduction target of 80% and five-year carbon budgets (Houses of the Oireachtas 2010b).

The context in which the government was operating was changed fundamentally by the global financial crisis, which precipitated a deep recession and collapse of the Irish banking sector. Irish Gross National Product fell by over 11% from 2007 to 2009. In the June 2009 local elections, Fianna Fail and the Green Party performed very poorly. Several independent TDs (members of parliament) defected from the government. As part of a renegotiation of the Programme for Government in October 2009, the Green Party secured a commitment to enact a climate change law, something that had not been included explicitly in the 2007 Programme for Government.

Alongside the deteriorating economic and political context, there was also growing pressure on governments around the world to be seen to be taking domestic action on climate change in the run-up to the COP15 UN climate conference in Copenhagen in December 2009. The Green Party used this rationale to progress the climate law domestically (Interview 3). Green Party environment minister John Gormley published a 'Framework for Climate Change Bill 2010' while the COP15 conference was taking place, and pledged to publish a full version of the bill before the end of 2010. This framework included provision for a 2050 GHG reduction target of 80% relative to 1990 and a requirement to reduce GHG emissions by 3% annually up to 2020.

As the Green Party worked towards publishing a draft climate law, the issue of quantified mitigation targets became one of the key sticking points. During 2010, business and farming lobby groups mounted a sustained attack on the inclusion of any targets in the climate bill beyond those already committed to by the state under EU and international obligations. The Irish Business and Employers' Confederation (IBEC) and the Irish Farmers' Association were particularly active in this regard (Interviews 1, 2, 3 & 7).

There was also significant resistance to the Green Party proposal across the civil service. The Departments of Agriculture, Finance, and Taoiseach were strongest in their opposition, with the Office of the Attorney General also strongly opposed. Even the Department of Environment was unsupportive (Interviews 2 & 6). The principal concern of civil servants was around the question of justiciability – whether failure to achieve targets specified in a climate law could result in a legal challenge to the government (Interview 4). As one senior civil servant involved in the process put it, 'This [inclusion of targets] was something that bothered everyone across the system' (Interview 6).

The broader economic and political context continued to deteriorate. The government was forced to request a bailout of EUR 85 billion from the EU and the International Monetary Fund in November 2010. Against this backdrop, environment minister John Gormley published the Climate Change Response Bill in December 2010 (Houses of the Oireachtas 2010a). The bill contained GHG reduction targets of 40% by 2030 and 80% by 2050. It also set out a requirement to reduce emissions by an average of 2.5% per annum from 2008 to 2020. However, against an increasingly turbulent backdrop, on 23 January 2011 the Green Party withdrew from government, resulting in a general election, and the Climate Change Response Bill did not progress further through the legislative process.

The efforts of the Green Party to develop climate legislation during its time in government, although unsuccessful, did push climate change legislation up the political agenda. In contrast to the 2007 election in which no political party contained an explicit commitment to climate legislation in its manifesto, the two parties that formed a coalition government after the 2011 election – the broadly Christian Democratic Fine Gael and the Labour

Party – committed to enacting a climate law in their election manifestos. Their programme for government committed to 'publish a Climate Change Bill which will provide certainty surrounding government policy and provide a clear pathway for emissions reductions, in line with negotiated EU 2020 targets' (Department of An Taoiseach 2011, p. 59). However, given the turbulent economic circumstances, climate legislation was not high on the agenda of the new government.

The first step taken by the new government was to publish a 'Review of National Climate Policy' in November 2011 (DELCG 2011). Significantly, this review made no mention of a climate law. In January 2012, Fine Gael environment minister Phil Hogan announced a roadmap for climate policy that committed to publishing a draft bill in Q4 of 2012 and enacting a climate law legislation by Q3/4 of 2013 (DELCG 2012). An outline of a new climate law, entitled the 'Climate Action and Low Carbon Development Bill', was published by Minister Hogan in February 2013 (DELCG 2013).

Among the two coalition partners, it was the Labour Party that was more proactive in seeking to progress climate legislation. However, in doing so the Labour Party faced significant resistance from civil servants, including officials in the Department of the Environment. According to one Labour Party official, 'Reluctance to have anything – any date, any target – was huge', with a key continuing concern among civil servants being the issue of justiciability (Interview 4). The new draft legislation shared some features with the draft bill that the Green Party had developed towards the end of its time in government, but there were also significant differences. The most important of these was the absence of any numerical targets for GHG emission reduction, in contrast to the previous bill that had included numerical targets for 2030 and 2050. The new draft bill committed to pursuing and achieving 'transition to a low carbon, climate resilient and environmentally sustainable economy in the period up to and including the year 2050'.

The draft climate law went through parliamentary scrutiny in mid-2013. According to an NGO representative involved in the process, NGOs and supportive backbench TDs took the view at this stage that the prospect of including numerical targets in the legislation was slim. Instead, they focused their attention on the design of an 'Expert Advisory Council' that was to be established under the legislation. NGO campaigners began to use the Irish Fiscal Advisory Council as a model that could be emulated in the climate legislation (Interview 2). This body had been established under the Fiscal Responsibility Act 2012 in response to the acknowledged failures of economic governance in Ireland in the lead-up to the economic crisis.

Nearly a year later, Minister Hogan published a revised outline of the legislation in April 2014 (DELCG 2014). Absent from this revised version again was any mention of long term or medium term emissions targets. However, the government published simultaneously a 'National Policy Position' on climate change that defined a long-term mitigation goal of 'an aggregate reduction in carbon dioxide (CO_2) emissions of at least 80% (compared to 1990 levels) by 2050 across the electricity generation, built environment and transport sectors; and in parallel, an approach to carbon neutrality in the agriculture and land-use sector, including forestry, which does not compromise capacity for sustainable food production' (Government of Ireland 2014). This was a compromise that involved agreement by government of a definition of 'low-carbon', but that was not enshrined in legislation. However, the term 'carbon neutrality' has eluded definition (CCAC 2017).

In January 2015, Alan Kelly, a Labour Party TD who had been appointed environment minister following Phil Hogan's nomination as European Commissioner the previous summer, published the full draft climate bill. As was the case during 2009, external pressure on governments to be seen to be acting on climate change increased as the COP21 Paris climate change approached. According to an NGO representative, this added impetus to the legislative process (Interview 2). Throughout the latter stages, lobby groups representing business and farming interests did not lobby extensively because they had no major objections to its content (Interview 1, 4, 5 & 7). In response to the January 2015 bill, Irish Farmers' Association spokesperson Harold Kingston stated that '[t]he Government's commitment not to introduce divisive and unachievable sectoral targets has been delivered in this Bill and represents a significant move forward from positions taken by previous Governments' (Irish Farmers' Association 2015).

In these latter stages NGOs focused their campaigning on a small number of issues: inclusion of a 2050 target for emissions reduction; independence of the Expert Advisory Body to be established under the climate law and removal of ex-officio members who were seen by NGOs to dilute its independence; inclusion of the principle of 'climate justice' in the legislation; and amending the timescale specified in the climate law for adopting the first national mitigation plan (Stop Climate Chaos 2015). In May 2015, the SCC coalition published a report it had commissioned from UK-based environmental law NGO, Client Earth, which explicitly compared Ireland's draft law with similar legislation from other European jurisdictions and sought to highlight perceived deficiencies in Ireland's draft law (Client Earth 2015). In particular, the report compared the draft law to the UK CCA as well as to recent legislation in Norway, Finland and Denmark, and argued in favour of including quantitative targets in the Irish law as well as strengthening the independence of the planned Expert Advisory Group.

NGOs had some success. Environment minister Alan Kelly agreed to inclusion of the climate justice principle. Regarding the Expert Advisory Group, he agreed to inclusion of a specific provision that it 'shall be independent in the performance of its functions'. Its name was also changed to 'Climate Change Advisory Council'. A government press release issued at the time indicated that the Council would model 'the format of the Fiscal Advisory Council' (Irish Government News Service 2015).

In autumn 2015 there was widespread speculation about a general election, which in the end was not held until February 2016. Nonetheless, this speculation prompted NGOs to scale back their lobbying for fear that the bill would fail to be enacted in advance of a general election (Interview 2). The Climate Action and Low Carbon Development Act was passed by both houses of parliament with all-party consensus and was signed into law by the President on 10 December 2015.

To sum up the Irish case, the UK CCA served as an inspiration for NGO campaigners, who returned again and again to the UK model. The Green Party and Labour Party also looked to the UK example, but primarily at an early stage (Interviews 3 & 4). However, by the end of the legislative process there were significant differences between Ireland's climate law and the UK CCA. Most prominently, the Irish climate law contained neither long term nor medium term quantitative GHG emission reduction targets. Although the Climate Change Advisory Council established under the Irish climate law mirrors the UK Committee on Climate Change to some extent, there is one crucial difference. The UK Committee is tasked with making a first recommendation on the UK's statutory carbon budgets, to which the government must then respond, thus giving the Committee a right of initiation of sorts. Since the Irish Climate Act does not contain provision for carbon budgets, the Irish CCAC does not have the same right. Moreover, with respect to mitigation and adaptation planning, the CCAC responds to government policies rather than having the right to make the first proposal.

The evolution that took place in the design of Ireland's climate law from early proposals to final legislation was shaped to a very significant extent by lobbying efforts of agricultural and business lobby groups who were opposed to several key design elements in early versions of the proposed legislation, as well as by significant resistance on the part of the civil service. According to a senior civil servant, 'We searched for international experience but couldn't find it. The UK was the only game in town but we didn't like it [the UK Act]' (Interview 6). This view was echoed in the reflections of a representative of an NGO representative: '[t]he civil service never accepted the basic premise of a [climate] law, that you would adopt a target and then adopt a pathway [to achieve that target] … There was a reluctance [among the civil service] to have their hands tied. They wanted maximum room for manoeuvre' (Interview 2).

The Finnish Climate Change Act

In Finland, the Ministry of Employment and Economy has, historically, held responsibility for domestic climate change and energy policy. In contrast to many other countries, the Ministry of Environment takes the lead only in international climate negotiations and not domestic policy (Teräväinen 2010). Using an advocacy coalition framework, Gronow and Ylä-Anttila have argued that the Ministry of Employment and Economy lies at the centre of an economy-oriented advocacy coalition that also comprises industry, trade unions and the three biggest political parties. This advocacy is most influential on climate and energy policy (Gronow and Ylä-Anttila 2016). An alternative environmental NGO coalition is significantly less influential, and while they enjoy formal access to policy processes, they do not enjoy high levels of influence.

The Finnish Government has drawn up and submitted to parliament national climate and energy strategies periodically – in 2001, 2005, 2008, 2013, and 2016. According to Hildén (2011), these strategies have evolved from 'clerical' recognition of obligations under the Kyoto Protocol to more visionary statements on the need for fundamental societal change. In 2014, a long-term Energy and Climate Roadmap for 2050 was drawn up by a Parliamentary committee in consultation with stakeholders. It set out a pathway to reducing Finland's GHG emissions by 80–95% between 1990 and 2050 (MEE 2014).

As in Ireland, the path to enactment of a climate change law in Finland originated with the launch of a 'Big Ask' campaign, which took place in Finland in February 2008. This campaign involved 16 NGOs and communities and was coordinated by FOE Finland (FOE Europe 2008). In framing this campaign, NGOs drew significantly on the UK Big Ask Campaign: 'In the beginning we were looking at a UK campaign which was a huge inspiration for us' (Interview 14). Over the course of 2008, nearly 17,000 people wrote to parliament demanding a climate law, and 45 MPs supported the call for a climate law (FOE Europe 2008).

Following this campaign, the Finnish Parliament in September 2008 agreed to consider two bills proposing to introduce a climate change law (Parliamentary bills LA 74/2008vp and LA 75/2008vp). According to Utter (2013) and Pölönen (2014), these bills were inspired by the UK CCA and the Big Ask campaign. However, neither bill secured enough votes to proceed. Also in autumn 2008, the Government commissioned a study to examine the feasibility of introducing a climate law modelled on the UK CCA in Finland. This study concluded that although some specific aspects of the UK model would not be possible in Finland, there were no legal obstacles to a climate law as such (Utter 2013, p. 3). However, with the failure of both 2008 bills to secure enough votes to proceed, the idea of a climate law fell off the legislative agenda during the rest of the parliamentary term.

The prospects for enacting a climate law were boosted considerably by the outcome of the 2011 general election. The election saw the rise of The Finns party, which increased its number of seats in parliament from 5 to 39. The ensuing negotiations to form a government saw an eventual agreement between six parties, without the involvement of The Finns party, which had withdrawn from negotiations over its opposition to Finland's support for the EU's bailout policy. Among the six remaining parties in the negotiations, three had promised a climate law in their election manifestos: the Social Democrats, the Green League, and the Left Alliance. Among these, it was most centrally a focus for the Greens (Interview 12). The eventual government agreement included a commitment to consider introducing a climate law as well as a pledge to establish a climate panel.

Over the following two years, stakeholder meetings were convened to discuss plans for a climate law, and a number of studies were undertaken to assess the feasibility of a climate law. The first, by academics Ari Ekroos and Matias Warsta, included overviews of Austrian, German, Scottish, US, and UK climate change regulations, and concluded that the UK model would not be practical in a small country such as Finland with its relatively smaller administration. It also found that a body modelled along the lines of the UK Climate Change Committee would be constitutionally difficult in Finland due to the fact that no institution outside the parliamentary and executive system can have anything other than an advisory role (Pölönen 2014, p. 306)(Interview 15). The study put forward a model climate law that included a 2050 GHG emission reduction target of 80% but no shorter term carbon budgets. Reaction to this model law was divided. The Ministry of Employment and Economy as well as EK, the main industry lobby, opposed it (Interview 15). NGOs, meanwhile, used the proposal to formulate their own, stronger model law which included a 2050 target of 95% reduction, a requirement for 5% annual emissions reductions, as well as a future review of the 2050 target (Interview 12).

In 2012, the Climate Panel, which had been established by government the previous year on a non-statutory basis, published its own report arguing for the necessity of such a climate law, and comparing different possible options including the UK CCA as well as the model law drafted by Ekroos and Warsta. The Climate Panel also found that the UK model would be difficult to implement in Finland because of differences in legal and administrative systems (Interview 10). Nonetheless, in contrast to Ekroos and Warsta, the Climate Panel report advocated the inclusion of short term as well as long term targets for mitigation and adaptation (Finnish Climate Panel 2012).

Early 2013 saw a ministerial decision that a climate law would be enacted before the end of the government's term in office in 2015 (Pölönen 2014, pp. 306–307). A preparatory committee at official level was formed to develop a draft climate law, chaired by the Environment Ministry and with representation from five ministries (Employment and Economy; Agriculture and Forestry; Traffic and Communications; Justice; Finance) and the Climate Panel. Important battles over the shape of the proposed climate law played out over the following year.

NGOs pushed three goals in their lobbying and campaigning: an ambitious long term target; a carbon budget approach, modelled on the UK CCA; and coverage of all GHG emissions, not just those outside the ETS sector (Interview 14). However, as time progressed, NGOs scaled back their demands for fear that a climate law might not be enacted at all (Interviews 13 and 14). According to one NGO representative, 'After government formation, it became obvious that if we combined level of ambition and the tool [a climate law], it would be a difficult fight. We had to leave the level of ambition for a different fight' (Interview 14).

The principal aim of industry lobbies, by contrast, was to keep energy prices low for industrial users. To this end, they sought to ensure that a climate law would not impact upon businesses, and were opposed to a carbon budget framework modelled on the UK CCA as well as any unilateral climate action by Finland (Interviews 8 and 11). Among relevant government ministries, the Environment Ministry was supportive of a climate law, but the Employment and Economy, Agriculture and Forestry, and Finance ministries were opposed (Interviews 13 and 15).

During the initial stages of framing the climate law, environmental NGOs along with Environment Minister and leader of the Green League, Ville Niinistö, pushed for a UK-inspired carbon budget model. However, the Green Party recognised early on that this would not be possible politically (Interviews 9 and 17). NGOs also pushed for the draft law to cover the whole Finnish economy. However, it quickly became clear that if a climate law were to be passed, it would not be possible to include the ETS sector because of strong opposition from industry groups in particular, as well as some government ministries (Interviews 11, 16 and 17).

There was a more substantial battle over the question of a 2050 target. Industry groups as well as the Ministry of Employment and Economy were opposed to such a target, while the Environment Ministry and NGOs favoured inclusion (Interview 16). Indeed, in the words of an NGO representative 'At what point would the bill become so weak that we wouldn't call it a success? The 2050 goal was key ... This became our final red line' (Interview 14). The Ministry of Employment and Economy did not achieve their aim of excluding a 2050 target, but secured inclusion of a stipulation that the 2050 target would be revisited in the context of an international agreement that set forth a different target (Interview 16).

The Environment Ministry and the Ministry of Employment and Economy disagreed over which ministry would have responsibility for policy planning under the proposed law (Interview 11). Under pre-existing arrangements, the Ministry of Employment and Economy had responsibility for drawing up periodic (once per parliamentary term) national climate and energy strategies. They were concerned about ceding control of this strategy planning process (Interview 14). Under the compromise reached, the Climate Change Act gives the Ministry of Employment and Economy responsibility for elaborating long-term climate change policy plans (to be issued every ten years), and gives responsibility for medium-term climate change policy plans (once per parliamentary term) to the Environment Ministry. In an unusual twist, however, the pre-existing climate and energy strategy process was retained by the Ministry of Employment and Economy, meaning that two shorter-term (once per parliamentary term) planning processes run side by side, each with a different ministry in charge, although with an agreement to work from the same set of scenarios and statistics (Interview 16).

The government published a draft of the bill in February 2014, with stakeholders given a chance to respond. Out of 69 responses received, 30 were clearly in favour while 19 were clearly against (Pölönen 2014, p. 307). On 5 June 2014, the government issued to parliament a proposal for a new climate change act (HE 82/2014). It was discussed by four parliamentary committees but there were no textual amendments at this stage. In September 2014 the Green League withdrew from the government coalition in protest over approval of a new nuclear power plant. However, this action did not derail the climate law, which was approved by Parliament on 6 March 2015 – a week before Parliament dissolved for a general election - based on a report by the parliament's environment committee. It was passed by 150 to 33 votes with near cross-party consensus – the Finns Party being the only political party to vote against it. The law entered into force on 1 June 2015.

To summarise, the UK CCA served as a reference point in the framing of Finland's climate law, somewhat more so than in Ireland. As with the Irish case, the NGO advocacy effort commenced with the 'Big Ask' campaign. Once the UK law took shape, this became a key reference point for NGO campaigning. Several studies also used the UK Act as a reference point, including the Government-commissioned Ekroos and Warsta report, and the Climate Panel's study, both of which found that the UK model would not be practical, at least not in its entirety, for Finland. Moreover, powerful ministries within the civil service were opposed to introducing a climate law, as were strong societal interests including business lobby groups. The 2011 general election result and the resulting complicated coalition negotiations provided the Green Party with an opportunity to put a climate law on the political agenda, but deep societal conflict resulted in a compromise climate law that resembled the UK Act only in very general terms, but without short or medium term climate targets or budgets, and with a relatively weak role for the Climate Panel.

Discussion and conclusion

I have explored whether the development of climate laws in Ireland and Finland constitutes policy diffusion from the UK, and tested two propositions: first, that international pressure to act on climate change pushes policymakers to develop climate policies that draw on experiences from other jurisdictions; and second, that interest group mobilisation and political competition over climate policy will cause states to adopt weaker, more symbolic versions of pioneering policies in other jurisdictions.

In respect of the first proposition, both cases witnessed clear evidence of the UK CCA being a source of inspiration in the early stages of the process, driven by international pressure to be seen to be acting on climate change. The lead-up first to COP15, and later COP21, created conditions in which policymakers wanted to be seen to be doing something. NGOs in Ireland and Finland used the UK Act as a model in their campaigning and lobbying, and mirrored the successful UK 'Big Ask' campaign in their national campaigns for climate legislation. In both cases, these campaigns were successful in putting the idea of a climate law on the political agenda, a cause also championed in both cases by Green Parties.

Regarding the second proposition, both cases witnessed strong societal conflict that led to weaker climate laws. Domestic interest groups mobilised to remove from legislative proposals the most pioneering and ambitious parts of the UK model, with the preferences of more powerful actors trumping NGO voices pushing for stronger legislation. In this respect, two key elements that facilitated passage of the UK CCA were absent in Ireland and Finland: support from large elements of business, and cross-party consensus on the need for strong climate legislation (Carter and Jacobs 2014).

In the Irish case, the Green Party's climate bill introduced at the end of 2010 included targets for 2020, 2030 and 2050, but did not replicate the 'carbon budget' element of the UK model. By the time the climate law idea was revived by the Fine Gael/Labour government in 2013, all targets – even for 2050 – had been removed. Powerful domestic lobby groups, including those representing business and agricultural interests, mobilised a successful campaign against a climate law modelled on the UK Act.

This echoes the findings of Wagner and Ylä-Anttila (2018), who find that the preferences of powerful economic actors saw their preferences reflected in the final version of the law. Moreover, the civil service – including the environment ministry in the earlier stages – was also opposed. In the face of this opposition, NGOs reined back their efforts out of fear that even a scaled-back law might not be passed.

The Finnish case shares many similarities with the Irish case. While the initial 'Big Ask' campaign in 2008 was unsuccessful, the idea of a climate law was prioritised by the Green League in government formation negotiations in 2011. Once detailed negotiations on the shape of a climate law got underway, business interests aligned with the powerful Ministry of Employment and Economy to lobby against a UK-inspired carbon budget model, which was ruled out at an early stage. However, this coalition did not succeed in its aim of excluding a 2050 target, which marks the Finnish case out from the Irish case.

This is not simply a case of differing contexts leading to different legislative outcomes that are similarly effective. Although both climate laws have created more systematic planning systems around climate policy, both have also been the subject of significant criticism. In the Irish case, this has been driven by the country's continued poor record on climate change. An illustration of the widely held view that Ireland's climate law is insufficient can be seen in the report from an all-party parliamentary committee on climate action, published in spring 2019 (Houses of the Oireachtas 2019). Responding to the climate change recommendations of the Irish Citizens' Assembly, this landmark report recommended that Ireland's climate law be significantly amended and remodelled closely on the UK CCA. Specifically, it recommended a system of five-yearly carbon budgets, and that the existing Climate Change Advisory Council be replaced with a new Climate Action Council (CAC) with enhanced functions and resources. The CAC would have a central advisory role in setting the five-yearly carbon budgets.

While the Finnish law also includes strong provisions for parliamentary and public scrutiny, it has been described as mainly procedural and largely symbolic, and also less ambitious than the preferences of the Finnish population (Schoenefeld *et al.* 2017). As Table 1 illustrates, the most pioneering aspects of the UK CCA were *not* transferred to Ireland and Finland in the process of diffusion. The result in both cases was climate laws that were less effective in terms of creating the framework for robust policy responses to climate change. In the Irish case, moreover, this outcome has led to renewed calls for Irish climate governance to be modelled on the UK CCA.

I have analysed the case of two later adopters of climate laws and showed clearly that, in these two cases, there was a significant paring back of the

Table 1. Principal elements of climate laws in the UK, Ireland and Finland.

	UK	Ireland	Finland
Long term target	80% by 2050	'Low carbon, climate resilient and environmentally sustainable economy' (no numerical target)	80% by 2050
Medium term targets	5 year carbon budgets	None	None
Planning obligations	• Carbon Plan (every five years, coinciding with carbon budgets) • UK Climate Change Risk Assessment (every 5 years) • National Adaptation Programme (every 5 years)	• National Mitigation Plan (every 5 years) • National Adaptation Strategy (to be updated every 5 years)	• Long-term climate change policy plan (every 10 years) • Medium-term climate change policy plan (once per electoral term) • National adaptation plan for climate change (every 10 years)
Reporting obligations	Annual report to Parliament by Committee on Climate Change	Annual Transition Statements to Parliament	Annual climate change report to Parliament
Advisory/ oversight committee	Committee on Climate Change	Climate Change Advisory Council	Climate Panel
Committee role	Clearly defined and extensive	Clearly defined but limited	Vaguely defined

Source: Author's compilation.

ambition displayed in the UK Climate Change Act. Future research could broaden the scope to consider examples of new climate laws that have been developed elsewhere in recent years. Future studies ought to study the dynamics of policy diffusion in respect of other types of climate policy instrument, and could also adopt a policy transfer or policy translation approach to examining whether and how the UK Act has influenced innovation in other countries directly. In the increasingly polycentric world of climate governance, it will become ever more important to understand the dynamics of such diffusion.

Note

1. Readers with a particular interest in the Irish case are directed to Torney (2017), which draws on the same empirical research as this section but develops the narrative at greater length.

Acknowledgments

An early draft of this article was presented at a workshop in June 2016 hosted by Dublin City University on 'Climate politics in small European states'. I thank participants in the workshop as well as Neil Carter, Conor Little, and two anonymous referees for helpful feedback, and Louise FitzGerald for editorial assistance. I am very grateful to Prof. Mikael Hildén and the Finnish Environment Institute for hosting me for a research visit in August 2016, and to the interviewees who shared their time and knowledge generously with me. The usual disclaimer applies.

Disclosure statement

No potential conflict of interest was reported by the author.

ORCID

Diarmuid Torney (iD) http://orcid.org/0000-0003-4156-9044

References

Benson, D. and Lorenzoni, I., 2014. Examining the scope for national lesson-drawing on climate governance. *The Political Quarterly*, 85 (2), 202–211. doi:10.1111/poqu.2014.85

Burck, J., *et al.*, 2018. *The climate change performance index: results 2019*. Berlin: Germanwatch.

Carter, N. and Childs, M., 2018. Friends of the Earth as a policy entrepreneur: 'The Big Ask' campaign for a UK climate change act. *Environmental Politics*, 27 (6), 994–1013. doi:10.1080/09644016.2017.1368151

Carter, N. and Jacobs, M., 2014. Explaining radical policy change: the case of climate change and energy policy under the British Labour government 2006–10. *Public Administration*, 92 (1), 125–141.

Cass, L.E., 2009. The symbolism of environmental policy. *In*: P.G. Harris, ed. *Environmental change and foreign policy: theory and practice*. London: Routledge, 41–56.

CCAC, 2017. *Annual review 2017*. Dublin: Climate Change Advisory Council.

Client Earth, 2015. *European lessons for Ireland's climate law*. London: Client Earth.

DECLG, 2011. *Review of national climate policy*. Dublin: Department of Environment Community and Local Government.

DECLG, 2012. *Hogan issues roadmap for climate policy and legislation*. Dublin: Department of Environment Community and Local Government.

DECLG, 2013. *General scheme of a climate action and low carbon development bill 2013*. Dublin: Department of Environment Community and Local Government.

DECLG, 2014. *General scheme of climate action and low carbon development bill 2014*. Dublin: Department of Environment Community and Local Government.

Department of An Taoiseach, 2011. Programme for government: statement of common purpose [online]. Available from: http://www.taoiseach.gov.ie/eng/Work_Of_The_Department/Programme_for_Government/Programme_for_Government_2011-2016.pdf [Accessed 11 July 2018].

EPA, 2018. *Ireland's Greenhouse Gas Emissions Projections, 2017–2035*. Wexford: Environmental Protection Agency.

Eurostat, 2018. Greenhouse gas emissions per capita [online]. Available from: http://ec.europa.eu/eurostat/tgm/table.do?tab=table&init=1&language=en&pcode=t2020_rd300&plugin=1 [Accessed 11 July 2018].

Falkner, R., 2012. Global environmentalism and the greening of international society. *International Affairs*, 88 (3), 503–522. doi:10.1111/inta.2012.88.issue-3

Fankhauser, S., Gennaioli, C., and Collins, M., 2016. Do international factors influence the passage of climate change legislation? *Climate Policy*, 16 (3), 318–331.

Finnish Climate Panel, 2012. *The Panel's report on the climate change act*. Helsinki: Finnish Climate Panel.

FOE Europe, 2008. New law to put finland at forefront of climate change fight [online]. Available from: http://www.foeeurope.org/press/2008/Sep26_New_law_to_put_Finland_at_forefront_of_climate_change_fight.html [Accessed 11 July 2018].

FOE Ireland, 2007. Climate protection bill 2007 [online]. Available from: http://www.foe.ie/download/pdf/climate_protection_bill.pdf [Accessed 11 July 2018].

Giddens, A., 2011. *The politics of climate change*. 2nded. Chichester: Wiley.

GLOBE, 2018. Global trends in climate change legislation and litigation: 2018 snapshot [online]. Available from: http://www.lse.ac.uk/GranthamInstitute/wp-content/uploads/2018/04/Global-trends-in-climate-change-legislation-and-litigation-2018-snapshot-3.pdf [Accessed 11 July 2018].

Government of Ireland, 2014. Climate action and low-carbon development: national policy position Ireland [online]. Available from: https://www.dccae.gov.ie/en-ie/climate-action/publications/Documents/5/National%20Climate%20Policy%20Position.pdf [Accessed 11 July 2018].

Green Party, 2007. Manifesto 2007 [online]. Available from: http://michaelpidgeon.com/manifestos/docs/green/Green%20Party%20GE%202007.pdf [Accessed 11 July 2018].

Gronow, A. and Ylä-Anttila, T., 2016. Cooptation of ENGOS or treadmill of production? Advocacy coalitions and climate change policy in Finland. *Policy Studies Journal*. Early View. doi:10.1111/psj.12185

Hildén, M., 2011. The evolution of climate policies: the role of learning and evaluations. *Journal of Cleaner Production*, 19 (6), 1798–1811.

Houses of the Oireachtas, 2007. Climate protection bill 2007 [online]. Available from: https://www.oireachtas.ie/documents/bills28/bills/2007/4207/b4207s.pdf [Accessed 11 July 2018].

Houses of the Oireachtas, 2010a. Climate change response bill 2010 [online]. Available from: https://www.oireachtas.ie/documents/bills28/bills/2010/6010/b6010d.pdf [Accessed 11 July 2018].

Houses of the Oireachtas, 2010b. *Second report on climate change law* Report of the Joint Committee on Climate Change and Energy Security. Dublin: Houses of the Oireachtas.

Houses of the Oireachtas, 2019. *Climate change: A cross-party consensus for action* Report of the Joint Committee on Climate Action. Dublin: Houses of the Oireachtas.

Irish Farmers' Association, 2015. Climate bill provides opportunity for the sustainable development of the agri-food sector [online]. Available from: http://www.ifa.ie/climate-bill-provides-opportunity-for-the-sustainable-development-of-the-agri-food-sector/#.V7oLdJh97tR [Accessed 11 July 2018].

Irish Government News Service, 2015. Minister Kelly proposes significant amendments to climate bill [online]. Available from: http://www.merrionstreet.ie/en/News-Room/Releases/Minister_Kelly_Proposes_Significant_Amendments_to_Climate_Bill.html [Accessed 11 July 2018].

Jensen, C.B. and Spoon, -J.-J., 2011. Testing the 'party matters' thesis: explaining progress towards Kyoto Protocol targets. *Political Studies*, 59, 99–115.

Klinger-Vidra, R. and Schleifer, P., 2014. Convergence more or less: why do practices vary as they diffuse? *International Studies Review*, 16 (2), 264–274.

Koch, M. and Fritz, M., 2014. Building the eco-social state: do welfare regimes matter? *Journal of Social Policy*, 43 (4), 679–703. doi:10.1017/S004727941400035X

Liefferink, D. and Wurzel, R.K.W., 2017. Environmental leaders and pioneers: agents of change? *Journal of European Public Policy*, 24 (7), 651–668.

Little, C., 2017. Intra-party policy entrepreneurship and party goals: the case of political parties' climate policy preferences in Ireland. *Irish Political Studies*, 32 (2), 199–223.

Lockwood, M., 2013. The political sustainability of climate policy: the case of the UK Climate Change Act. *Global Environmental Change*, 23, 1339–1348.

McGee, H., 2009. Climate change act to closely follow legislation in UK, says Ryan [online]. Available from: http://www.irishtimes.com/news/climate-change-act-to-closely-follow-legislation-in-uk-says-ryan-1.772671 [Accessed 11 July 2018].

MEE, 2014. *Energy and climate roadmap 2050*. Helsinki: Ministry of Employment and the Economy.

Newell, P. and Paterson, M., 2010. *Climate capitalism*. Cambridge: Cambridge University Press.

Newig, J., 2007. Symbolic environmental legislation and societal self-deception. *Environmental Politics*, 16 (2), 276–296. doi:10.1080/09644010701211783

Pölönen, I., 2014. The Finnish Climate Change Act: architecture, functions, and challenges. *Climate Law*, 4, 301–326. doi:10.1163/18786561-00404006

Schoenefeld, J.J., Hildén, M., and Mäkinen, K., 2017. The Finnish Climate Change Act: in line with what Finland's public wants? [online]. Available from: http://environmentaleurope.ideasoneurope.eu/2015/06/12/finnish-climate-change-act-line-finlands-public-wants/[Accessed 11 July 2018].

Simmons, B.A., Dobbin, F., and Garrett, G., 2006. Introduction: the international diffusion of liberalism. *International Organization*, 60 (4), 781–810.

Stop Climate Chaos, 2015. Briefing paper February 2015: the climate action and low carbon development bill 2015 and the recommendations of the Joint Committee on the Environment, Culture and the Gaeltacht [online]. Available from: https://www.trocaire.org/sites/default/files/pdfs/campaigns/scc-briefing-2015.pdf [Accessed 11 July 2018].

Teräväinen, T., 2010. Ecological modernisation and the politics of (un)sustainability in the Finnish climate policy debate. *In*: S. Simard, ed.. *Climate change and variability*. Rijeka: InTech, 409–426.

Torney, D., 2017. If at first you don't succeed: the development of climate change legislation in Ireland. *Irish Political Studies*, 32 (2), 247–267. doi:10.1080/07907184.2017.1299134

Utter, R., 2013. A climate change act: comments from a Finnish legal perspective. University of Eastern Finland Law School Legal Studies Research Papers. Paper No. 12. doi:10.2139/ssrn.2327533

Vogler, J., 2016. *Climate change in world politics*. Basingstoke: Palgrave.
Wagner, P. and Ylä-Anttila, T., 2018. Who got their way? Advocacy coalitions and the Irish climate change law. *Environmental Politics*, 27 (5), 872–891.

Appendix I. *Details of interviewees*

Irish interviews

Interview 1: Business representative
Interview 2: Environmental NGO representative
Interview 3: Green Party politician
Interview 4: Labour Party advisor
Interview 5: Senior civil servant
Interview 6: Senior civil servant
Interview 7: Farmers' representative

Finnish Interviews

Interview 8: Business representative
Interview 9: Member of Finnish Climate Panel
Interview 10: Academic
Interview 11: Business representative
Interview 12: Environmental NGO representative
Interview 13: Environmental NGO representative
Interview 14: Environmental NGO representative
Interview 15: Academic
Interview 16: Senior civil servant
Interview 17: Senior civil servant

Index